GCSE 9-1 geography

EDEXCEL B

Revision Guide

Series editor
Bob Digby

David Holmes
Rebecca Priest

Andy Slater
Kate Stockings
Rebecca Tudor

OXFORD
UNIVERSITY PRESS

OXFORD
UNIVERSITY PRESS

Great Clarendon Street, Oxford, OX2 6DP, United Kingdom

Oxford University Press is a department of the University of Oxford. It furthers the University's objective of excellence in research, scholarship, and education by publishing worldwide. Oxford is a registered trade mark of Oxford University Press in the UK and in certain other countries

© Oxford University Press 2019

Series editor: Bob Digby

Authors: David Holmes, Rebecca Priest, Andy Slater, Kate Stockings, Rebecca Tudor

The moral rights of the authors have been asserted

Database right of Oxford University Press (maker) 2019.

First published in 2019

British Library Cataloguing in Publication Data
Data available

978-0-19-843623-2

1 3 5 7 9 10 8 6 4 2

Paper used in the production of this book is a natural, recyclable product made from wood grown in sustainable forests. The manufacturing process conforms to the environmental regulations of the country of origin.

Printed in Great Britain by Bell and Bain Ltd., Glasgow

Acknowledgements

The publisher and authors would like to thank the following for permission to use photographs and other copyright material:

Cover: Revenant/Shutterstock; Pavel K/Shutterstock. **p8:** Bob Digby; **p9:** Stocktrek Images, Inc./Alamy Stock Photo; **p10:** Chris Wildt/Cartoonstock; **p16:** NASA; **p22:** Stefano Garau/Shutterstock; **p25:** With kind permission from the National Hurricane Centre; **p47:** Dinodia Photos/Alamy Stock Photo; **p54:** Stuart Boulton/Alamy Stock Photo; **p58(l):** Wikimedia Commons/Public Domain; **p58(r):** Universal Images Group North America LLC/Alamy Stock Photo; **p59:** Daniel Berehulak/Getty Images); **p60:** AFP/Getty Images; **p63:** Medicshots/Alamy Stock Photo; **p69(t):** Kevin Eaves/Shutterstock; **p69(b):** Ernie Janes/Alamy Stock Photo; **p70:** Flasshary; **p73(t):** Bob Digby; **p73(b):** Mike Page Photography; **p76:** Bob Digby; **p78:** Ordnance Survey © Crown copyright and Database rights 2019; **p79:** Bob Digby; **p80:** Gary Clarke/Alamy Stock Photo; **p81:** Bob Digby; **p82:** Neil Holmes Photography; **p86:** Matt Cardy/Getty Images; **p96:** High Level Photography Ltd; **p99(t):** MS Bretherton/Alamy Stock Photo; **p99(b), 101(t), 102(b):** Bob Digby; **p102:** Ordnance Survey © Crown copyright and Database rights 2019; **p104:** Angela Hampton Picture Library/Alamy Stock Photo; **p107(t), 107(b):** Bob Digby; **p108:** Ordnance Survey © Crown copyright and Database rights 2019; **p110, 111, 114, 115, 116, 119, 120, 121, 124, 125, 126:** David Holmes; **p129:** Bob Digby; **p133:** Oleg Znamenskiy/Shutterstock; **p144:** Rob Simmon and Jesse Allen, NASA's Earth Observatory; **p151(l):** Justin Sullivan/Getty Images; **p151(r):** David Fulford/Ashden/www.ashden.org/winners/biotech; **p152(l):** © Lu Guang/Greenpeace; **p152(r):** Phil Clarke Hill/Getty Images; **p159:** Eamon Mac Mahon/Associated Press; **p160:** Justin Kase zsixz/Alamy Stock Photo; **p161:** Courtesy of Toyota; **p167:** Prisma by Dukas Presseagentur GmbH/Alamy Stock Photo; **p169:** Victor St. John/Alamy Stock Photo.

Artwork by Mike Connor, Barking Dog Art, Simon Tegg, Q2A Media Services Inc., and Aptara Inc.

Every effort has been made to contact copyright holders of material reproduced in this book. Any omissions will be rectified in subsequent printings if notice is given to the publisher.

Contents

Contents

Component 3: People and environment issues – Making geographical decisions

Contents

 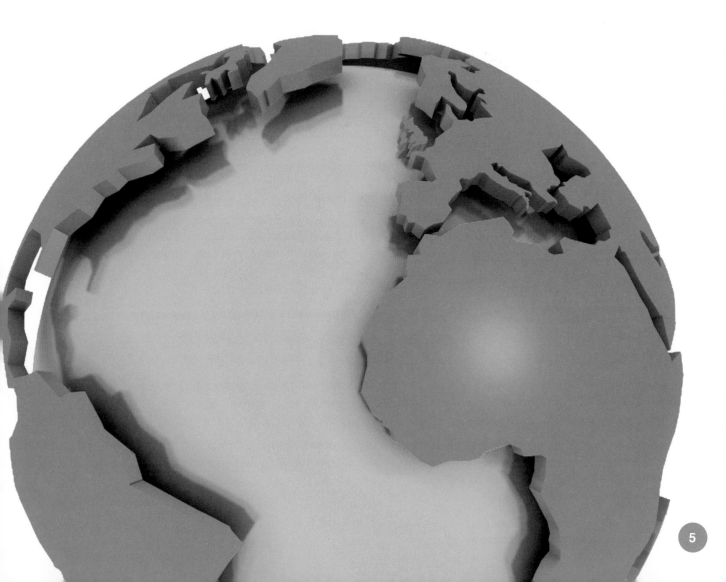
Guided answers are available on the Oxford Secondary Geography website: www.oxfordsecondary.co.uk/geography-answers

Please note this revision guide has not been written or endorsed by Pearson Edexcel. The answers and commentaries provided represent one interpretation only and other solutions may be appropriate.

Introduction: Helping you succeed

If you want to be successful in your exams, then you need to revise all you've learned in your GCSE course! That can seem daunting – but it's why this book has been written. It contains key revision points that you need to know to do well in the Edexcel GCSE Geography B specification exams.

Your revision guide

This book is designed to help you revise for your three Edexcel GCSE Geography B exam papers.

Each component is split into chapters. Each chapter has an introduction page which contains an outline of:

- the three exam papers you'll be taking
- the key ideas and content that form the specification.

Within each chapter are single-page units which contain the following features:

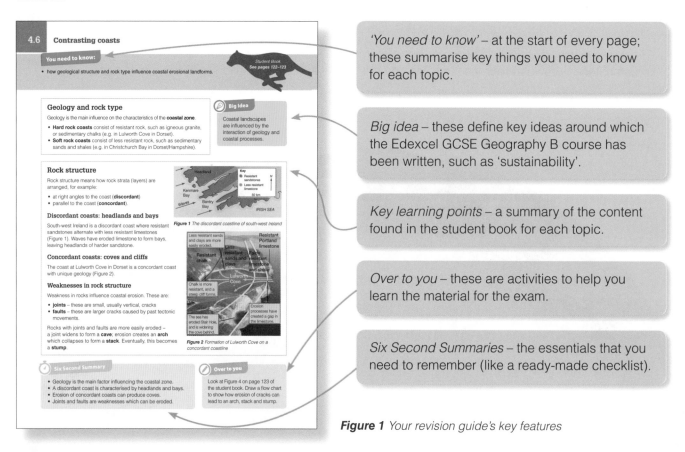

'You need to know' – at the start of every page; these summarise key things you need to know for each topic.

Big idea – these define key ideas around which the Edexcel GCSE Geography B course has been written, such as 'sustainability'.

Key learning points – a summary of the content found in the student book for each topic.

Over to you – these are activities to help you learn the material for the exam.

Six Second Summaries – the essentials that you need to remember (like a ready-made checklist).

Figure 1 Your revision guide's key features

Your revision

Each topic (1.1, 1.2 etc.) in this revision guide exactly matches the content for each topic in your Edexcel GCSE Geography B student book. Key content in each double page unit in the student book is summarised in a single page in the revision guide.

The Edexcel GCSE Geography B specification has three components. Each component contains topics. Each component is assessed by an exam (Paper 1, 2, 3) with sections for different topics, as follows:

Topics you need to learn

Component 1 Global geographical issues

This is assessed by Paper 1 in the exam. It consists of three sections, each containing different topics. You must do all questions in the paper.

- **Section A** includes Topic 1 Hazardous Earth, with questions on the global climate system, climate change, extreme weather (e.g. tropical cyclones) and tectonic hazards.

- **Section B** includes Topic 2 Development dynamics, with questions on global development and a case study of one of the world's emerging countries.

- **Section C** includes Topic 3 Challenges of an urbanising world, with questions on rapid urbanisation and global urban trends, and a case study of one of the world's megacities in either a developing or an emerging country.

In addition, there will be questions on geographical skills (e.g. how to interpret statistics, maps, diagrams or photos) in every topic.

Memory jogger for Paper 1!

- My case study of an emerging country was of
  _____

- My case study of a developing or an emerging country was of
  _____
  which is a developing / emerging country (delete one)

Component 2 UK geographical issues

This is assessed by Paper 2 in the exam. It consists of four sections, each on a different topic. You have a choice of fieldwork questions in Section C.

- **Section A** includes Topic 4 The UK's evolving physical landscape with questions on the UK's physical landscape, coastal change and conflict, and river processes and pressures.

- **Section B** includes Topic 5 The UK's evolving human landscape with questions on the UK's changing population and a case study of one major UK city.

- **Sections C1 and C2** include Topic 6: Geographical investigations, with questions on fieldwork in Section C1 on **either** Investigating coastal change and conflict **or** Investigating river processes and pressures, and in Section C2 on **either** Investigating dynamic urban areas or Investigating changing rural areas.

As with Paper 1, you will also have questions on geographical skills in every topic.

Memory jogger for Paper 2!

- My case study of a major UK city was of
  _____

- My physical fieldwork was on rivers / coasts (delete one) and we collected data on _____
  at (name of place)
  _____

- My human fieldwork was on cities / rural areas (delete one) and we collected data on
  _____
  at (name of place)
  _____

Component 3 Making geographical decisions

This is assessed by Paper 3 in the exam. It consists of an unseen Resource Booklet with the following sections:

- **Section A** includes Topic 7 People and the biosphere.

- **Section B** includes Topic 8 Forests under threat (including examples from rainforests and the taiga).

- **Section C** includes Topic 9 Consuming energy resources.

- **Section D** includes a geographical decision-making question.

Memory jogger for Paper 3!

- My case study of rainforest was of
  _____

- My case study of taiga was of
  _____
  _____

Know your key words!

In your GCSE course you've been learning to 'think like a geographer' – about the processes affecting the Earth's surface, and how people affect the natural environment. Now, you must learn to write like a geographer, which means knowing key geographical words.

Key words help you to:

- understand a question (e.g. explaining how a process occurs)
- identify features in diagrams or photos, like Figure 2
- use key words in your answers. Answers that use key words earn more marks than those that don't.

- Beach
- Larger sediment particles
- Finer beach material
- Evidence of slumping
- Cliff face

Figure 2 *A coastline in Cornwall – which of these key terms can you recognise?*

Many words used in geography are used in day-to-day life, such as 'population' or 'beaches'. But you only ever meet some key words in certain topics (e.g. precipitation). Many are key to the subject – like the words in Figure 3 below.

Do you know these key words?

In the left-hand column are definitions of terms you should know. Write them in the right-hand column 1 to 5, then check the answers on page 9.

Figure 3 *Which key words are these?*

Description	Geographical term
The rock type of an area, or the study of rocks	1
People coming to live in a country from overseas	2
Gases which warm the atmosphere	3
Specialist service and technical industries	4
The total value of goods and services in a country in a year	5

Mmmm …. mnemonics!

Learning key words and processes can be hard. Even something like spit formation has ten steps to remember (see below). However, **mnemonics** can help – it's a way of developing a system to help your memory.

How Mnemonics work

1. Try this way of learning processes that form a coastal spit:

   a) Winds blow at an **angle**

   b) Waves break – **swash**

   c) Sand moves **up** the beach

   d) Water moves down the beach – **backwash**

   e) Sand moves down the beach – forming a **zigzag**

   f) This is called **longshore** drift

   g) Sand reaches an **estuary**

   h) Beach forms a **spit**

   i) River currents form a **hook**

   j) **Mudflats** form behind the spit

2. Look at the words in bold – then take the first letter – **A**ngle, **S**wash, **U**p, **B**ackwash, **Z**igzag, **L**ongshore, **E**stuary, **S**pit, **H**ook, and **M**udflats.

3. Write the letters in a list, then create a sentence from the first letters to help you remember – it can be as daft as you like!

4. Learn the sentence!

Over to you

Try making mnemonics for:

a) the erosion processes on coasts (section 4.11), or in rivers (section 4.15)

b) key words in any **one** other topic that you've revised.

Revising examples and case studies

Hazards is one of the most popular topics in geography. Hazard events such as the Haiti earthquake of 2010 (Figure 4) are among the many examples and case studies that you need to revise for the exam.

- You can use this approach for any hazard – for example, a flood, volcanic eruption, or tropical cyclone.
- You can adapt this approach to revise **any** example or case study.

Figure 4 *The Haiti earthquake in 2010*

1

Build up a factfile

Revising case studies can be hard because there's a lot of detail to remember. Start by building up a basic factfile, such as:

- Location – where is it? Can you locate it in an atlas?
- What kind of hazard was it?
- When did it occur? Date and time?
- Was it a single event or one of several?
- Describe what happened e.g. for an earthquake – what date, what time of day, how strong was it, where was the epicentre?
- What were its causes? Is it on a destructive plate margin?

2

Know its impacts

The impacts – or effects – of many hazard events can be significant. What were its short-, medium- and long-term impacts? Be clear that you know what this means:

- Short-term – within the first month or so
- Medium-term – within six months
- Longer-term – anything longer than six months, maybe even years.

Next, classify these into economic, social or environmental impacts.

- Economic impacts (related to money, e.g. jobs, businesses, trade, costs).
- Social impacts (about people, health, and housing).
- Environmental impacts (about changes to the surrounding landscape).

You can now list these impacts and classify them, using the grid in Figure 5.

Impact	Immediate/short-term	Medium-term	Long-term
Economic			
Social			
Environmental			

Figure 5 *A table for classifying the impacts of a hazard event such as a volcanic eruption or an earthquake*

 Over to you

Think about how you could adapt a) the factfile, b) the table of impacts, to other topics. Consider these possibilities:

1) A study of a developing or an emerging country. What sort of factfile would you produce?

2) A study of a megacity in a developing or emerging country. How would you build up notes on reasons for its growth or problems it is trying to solve?

3) A study of a major UK city (e.g. Bristol). How would you build up notes on ways in which it is trying to become more sustainable?

Answers to Figure 3, page 8

1 Geology

2 Immigration (or immigrants)

3 Greenhouse gases

4 Quaternary industries

5 GDP

How to get the best marks possible in your exams? To reach your best standard of writing on longer answers, follow the advice below.

1

Plan your answer

It helps to organise your thoughts if you plan your answer. You don't need a long plan – just something that takes you 30 seconds to jot down. Some people use a spider diagram, others just make a short list. It helps you to get the order of the answer right, and makes sure you don't forget what to write. One example of a plan is given in Figure 6.

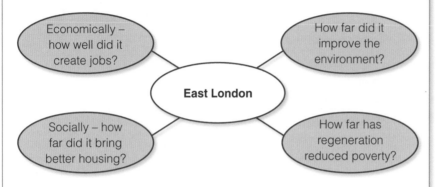

Figure 6 *An example of a plan for an 8-mark question:* Assess the success of attempts to regenerate parts of one major UK city. (8 marks)
Hint: *You could use page 101 in this Revision Guide to help you.*

"We prefer to call this test 'multiple choice,' not 'multiple guess.'"

Figure 7 *The dangers of not revising!*

 Over to you

In the box to the left, draw a plan like the one in Figure 6 for the question:

Assess the ways in which one major UK city is trying to become more sustainable. (8 marks)

Hint: you could use page 103 in this Revision Guide to help you.

Improving your exam answers

Use key words

Below are two exam answers where candidates A and B were supposed to use key words.

- Candidate A answer gained only 2 marks out of 4.
- Candidate B answer gained only 3 marks out of 8.

In each case, the candidate knew the general idea but didn't use any key words.

Over to you

1. Below are two candidate answers from an exam. To the right of each answer are key geographical words. Replace each word in bold in the answer with one of the key geographical words.

2. Write the improved answer in the space to get full marks.

Candidate answer A

Describe two ways in which human activity leads to climate change.
(4 marks)

People are using more and more cars, and the gases that **are given off** go up into **the air** and **damage** the air we breathe and **make it warmer**. By **cutting down trees**, less oxygen is produced and there is **more of other gases** in the air making the **weather** warmer.

_____

_____

_____

_____

_____

_____

_____

Key geographical words to improve the answer:
atmosphere
carbon dioxide
climate
emitted
deforestation
increase the greenhouse effect
pollute

Candidate answer B

Assess the ways in which human actions can increase the risk of flooding.
(8 marks)

Human actions can increase flood risk. First, **cutting down trees** so there is grass for cattle to **feed on** and fuel means that there are no more trees to **break the fall of rain** so it gets to the soil straight away. More rain gets to the river by **flowing over the land** and the river **fills up** very quickly and **overflows**. Tree roots also bind the soil together, and if the trees are **cut down** then the soil **wears away**. This soil then gets into the river and gets **carried along** and **dropped** on the **bottom of the river** which raises it and reduces **how much water the river can hold**.

_____

_____

_____

_____

_____

_____

_____

Key geographical words to improve the answer:
capacity
deforestation
deforested
deposited
erodes
floods
graze
intercept
reaches capacity
river bed
surface runoff
transported

Topic 1
Hazardous Earth

Your exam

- Topic 1 Hazardous Earth makes up Section A in Paper 1, Global geographical issues.

- Paper 1 is a 90-minute written exam and makes up 37.5% of your final grade. The whole paper carries 94 marks (including 4 marks for SPaG) – questions on Topic 1 will carry 30 marks.

- You have to answer all questions in Paper 1. Section B contains questions on Development dynamics (pages 34–49), and Section C on Challenges of an urbanising world (pages 50–63).

Tick these boxes to build a record of your revision

Your revision checklist

Spec Key Idea	Detailed content that you should know	1	2	3
1.1 The atmosphere operates as a system which transfers heat around the Earth	• The global atmospheric circulation, and ocean currents			
	• How this determines the location of arid and high rainfall areas			
1.2 Climate has changed in the past through natural causes	• Natural causes of climate change			
	• Evidence for natural climate change and how it is used to reconstruct the past			
1.3 Global climate is now changing as a result of human activity	• How human activities cause the enhanced greenhouse effect			
	• Consequences on people			
	• Future projections for climate change and sea level rise			
1.4 Tropical cyclones are caused by particular meteorological conditions	• Characteristics and global distribution of tropical cyclones			
	• Why tropical cyclones develop, intensify, and die out			
1.5 Tropical cyclones present major natural hazards to people and places	• Hazards of tropical cyclones and their impact on people and environments			
	• Why some countries are more vulnerable than others to the impacts of tropical cyclones			
1.6 The impacts of tropical cyclones are linked to a country's ability to prepare and respond to them	• How countries can prepare for, and respond to, tropical cyclones			
	• Preparation and response in two contrasting countries			
1.7 Earth's layered structure and physical properties is key to plate tectonics	• Earth's structure and physical properties			
	• How the core's internal heat source generates convection and plate motion			
1.8 There are different plate boundaries, each with characteristic volcanic and earthquake hazards	• Distribution and characteristics of the three plate boundary types and hotspots			
	• Causes of contrasting volcanic and earthquake hazards, including tsunami			
1.9 Tectonic hazards affect people, and are managed differently	• Impacts of earthquakes or volcanoes			
	• Managing volcanic or earthquake hazards			

Student Book
See pages 8–9

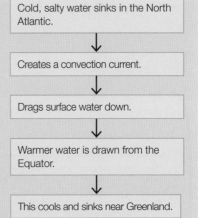

You need to know:

- how the sun heats different parts of the Earth
- how pressure differences and ocean currents affect temperatures.

Keeping the Earth habitable

Figure 1 shows how the sun's energy (**solar insolation**) varies around the Earth.

- This heat is redistributed by **pressure differences** and **ocean currents**.
- If it wasn't redistributed, the tropics would be hotter, and the polar regions even colder than they are.

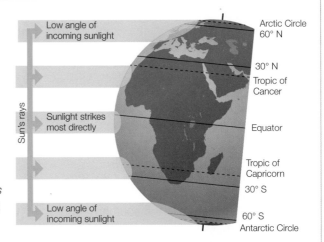

Figure 1 *The angle of the sun's rays makes solar insolation very intense at the Equator, but dispersed at the Poles*

Labels on diagram:
- Low angle of incoming sunlight
- Sunlight strikes most directly
- Low angle of incoming sunlight
- Sun's rays
- Arctic Circle 60° N
- 30° N, Tropic of Cancer
- Equator
- Tropic of Capricorn
- 30° S
- 60° S, Antarctic Circle

Pressure differences

Land and sea heat up differently.

Land:

- heats quickly in summer and cools quickly in winter.
- heats the air above, which expands, becomes lighter and rises.
- forms areas of low pressure in the summer and high pressure in winter.

Sea:

- takes longer to heat and cool. So the air above is dense and cool in the summer.
- forms areas of high pressure in summer, and low pressure in winter.

Wind is created when air moves from high to low pressure.

- In January in the southern hemisphere, low pressure forms over South America, Africa and Australia as they are warmer than the oceans.
- Because the oceans are cooler, they form areas of high pressure.
- The opposite happens at the same time in the northern hemisphere.

Ocean currents

Figure 2 explains how a warm **ocean current** called the Gulf Stream increases temperatures in Europe.

Wind moves warm water from the Gulf of Mexico north-eastwards towards Europe.

Figure 2 *The Gulf Stream is one of several ocean currents that move heat around the globe*

Flow chart:
- Cold, salty water sinks in the North Atlantic.
- ↓
- Creates a convection current.
- ↓
- Drags surface water down.
- ↓
- Warmer water is drawn from the Equator.
- ↓
- This cools and sinks near Greenland.
- ↓
- It flows south to the Equator and is warmed again.

 Six Second Summary

- The amount of heat from the sun's rays varies around the Earth.
- Land and sea heat up differently, creating areas of high and low pressure.
- Heat is distributed around the Earth due to pressure differences (wind) and ocean currents.

Over to you

Summarise how pressure differences affect global temperatures in two sentences. Then, in one sentence, then five words, then two words. Do the same for ocean currents!

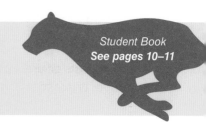

Student Book
See pages 10–11

You need to know:

- how the global circulation model brings rainfall to Africa
- the role of the ITCZ and the Hadley Cell in the global circulation model.

The ITCZ

The Inter-Tropical Convergence Zone (ITCZ) is an area of low pressure. It stretches around the Earth and brings rain.

- In June, when the sun is over the Tropic of Cancer, the ITCZ moves north, bringing rain to the northern hemisphere (e.g. West Africa).
- In December, when the sun is over the Tropic of Capricorn, the ITCZ moves south, bringing rain to the southern hemisphere (e.g. southern Africa).

So the ITCZ brings rain to different regions of Africa at different times of the year (Figure 1).

What causes the ITCZ?

The ITCZ:

- is part of the **global circulation model** (Section 1.3)
- forms within the **tropics**
- forms where two air masses **converge** (meet)
- is created within the **Hadley Cell** – the largest of the three air 'cells' in the global circulation model.

How does the Hadley Cell work?

The Hadley Cell:

- is caused by heating and cooling
- creates low and high pressure systems (Section 1.1)
- consists of two parts, one either side of the Equator
- both parts move together as the sun moves overhead seasonally.

Figure 1 below shows how the Hadley Cell brings rain to West Africa in July.

A – The land warms the air. It expands, becomes lighter and rises. There is low pressure over the Southern Sahara.

B – Cooler, denser, high pressure air forms as the sea is cooler than the land.

C – The 'Trade Winds' blow from high to low pressure towards the Southern Sahara. They carry moisture to West Africa.

D – High pressure forms over the Atlantic Ocean because the sea is cooler than the land in the northern hemisphere summer.

E – Trade Winds blow south, drawn by low pressure at the ITCZ. They converge with the winds from C, rise and cool. Water vapour condenses to make rain.

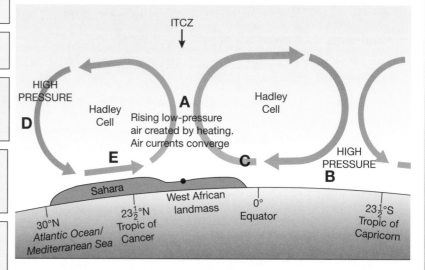

Figure 1 *How the Hadley Cell forms the ITCZ and brings rain to West Africa in July*

  Six Second Summary

- The ITCZ is formed by the Hadley Cell, part of the global circulation model.
- The ITCZ is where the Trade Winds of the Hadley Cell converge, rise and bring rain.
- The ITCZ moves between the Tropics, bringing rain to West Africa in July.

Over to you

Make a copy of the Hadley Cells in Figure 1. Annotate them to show how they bring rain to West Africa.

*Student Book
See pages 12–13*

You need to know:

- how the global circulation model affects areas of high and low rainfall.

Completing the cycle

As well as bringing rainy seasons to West Africa in summer, the Hadley Cell also brings the dry season in winter. Section 1.2 looked at the global circulation model in July. Figure 1 shows what happens in January.

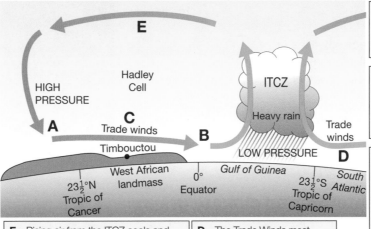

A – It is winter in the northern hemisphere. There is cool, dense high pressure over North Africa.

B – The southern hemisphere warms as the sun is over the Tropic of Capricorn. This forms low pressure.

C – The 'Trade Winds' blow from high to low pressure towards the Southern Sahara. They carry dry air from the Sahara over West Africa, causing its dry season.

E – Rising air from the ITCZ cools and becomes denser. It spreads out as is reaches high in the atmosphere and falls as high pressure air, returning to **A**.

D – The Trade Winds meet others from the southern hemisphere to form the ITCZ. It rains in southern Africa.

Figure 1 The Hadley Cell in January

The global circulation model

Two other cells complete the global circulation model – the **Ferrel Cell** (30°-60°N and S) and the **Polar Cell** (60°-90°N and S). They are found in each hemisphere (Figure 2).

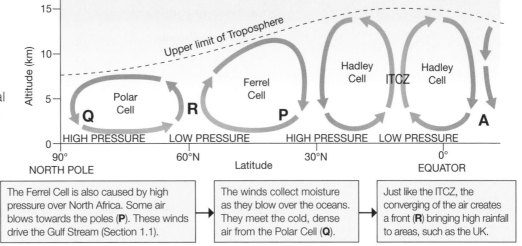

The Ferrel Cell is also caused by high pressure over North Africa. Some air blows towards the poles (**P**). These winds drive the Gulf Stream (Section 1.1).

The winds collect moisture as they blow over the oceans. They meet the cold, dense air from the Polar Cell (**Q**).

Just like the ITCZ, the converging of the air creates a front (**R**) bringing high rainfall to areas, such as the UK.

Figure 2 The three 'cells' of air in the global circulation model

Explaining global rainfall

The global circulation model affects climates around the world.

- In June, rains brought by the ITCZ don't reach the Sahara, so it remains dry. In January, high pressure brings cool, dry air, so it is dry then too.
- Dry, dense air at the Poles forms **polar deserts**.

Six Second Summary

- The global circulation model affects the amount and location of global rainfall.
- The three air 'cells' create areas of high and low rainfall.

Over to you

Draw and label the Hadley Cell. Explain in bullet points how it is linked to the ITCZ.

Geographical skills: learning about climate

You need to know:

- how to use satellite images
- how to draw and interpret climate graphs
- how to track cyclones.

Student Book
See pages 14–15

1 Using satellite images

Satellite images (Figure 1) are:

- images taken from space satellites
- a form of geographical information systems (GIS)
- used to accurately predict weather.

They're like maps – to understand them, you need an outline map, north arrow and key.

The sea looks relatively dark

UK

Land looks lighter than the sea

Clouds appear light grey or white

Cloud showing the ITCZ

Cloud showing the ITCZ

Figure 1 *A satellite image showing the ITCZ over West Africa, June 2012*

2 Drawing and interpreting climate graphs

Climate graphs show:

- rainfall and temperature on the same graph
- temperature (in °C) as a line
- rainfall (in mm) as bars
- months on the x axis.

To draw a climate graph:

- begin your axes at 0
- prepare the rainfall scale by looking at the highest rainfall value
- prepare the temperature scale by looking at the highest temperature value
- plot each rainfall bar, 1 cm wide, beginning with January
- plot temperature data as points
- join the points.

To interpret a climate graph:

- calculate the **mean** annual temperature
- calculate the temperature **range**
- analyse the temperature **trends**
- calculate the **total** annual rainfall
- identify the **wettest** and **driest** months
- see how sharply the rainfall varies.

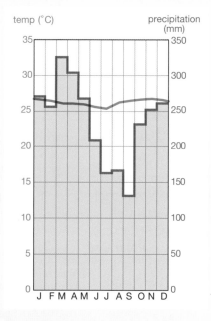

Figure 2 *A climate graph*

3 Tracking cyclones

Tracking cyclones:

- is vital for hazard prediction
- uses a series of points joined together to provide a map
- is used alongside satellite images, sea surface temperatures and cyclone history to predict which way a cyclone will move.

 **Six Second Summary**

- Satellite images taken from space help to predict weather.
- Climate graphs tell you about the annual rainfall and temperature of a location.
- Tracking cyclones helps to predict their movement.

Over to you

Using the instructions for interpretation listed above, annotate a copy of Figure 2.

The causes of climate change in the past

You need to know:

- the four main theories that explain why climate has changed in the past.

Student Book
See pages 16–17

The eruption theory

Large volcanic eruptions can alter Earth's climate for a few years.

- Eruptions produce ash and sulphur dioxide.
- These rise to the **stratosphere** and reflect some sunlight back into space, so the planet cools.
- In 1991, the eruption of Mount Pinatubo cooled the planet by 0.5°C for a year.

Asteroid collision theory

Big asteroid impacts can alter Earth's climate for 5–10 years.

- 1 km-sized asteroids hit Earth every 500 000 years.
- Millions of tonnes of ash and dust would be blasted into the atmosphere, blocking incoming sunlight and cooling the climate.

Sunspot theory

Sunspots are black areas on the sun's surface which appear when the sun is more active.

- Lots of sunspots means more solar energy is coming to Earth, e.g. during the Modern Warm Period.
- Cooler periods could have been caused by fewer sunspots.

Big Idea

The four causes of past climate change are all natural.

Orbital change theory

Long term changes in climate could be caused by the way the Earth orbits the sun (Figure 1).

- The Earth's orbit is sometimes circular, sometimes more oval.
- The Earth's tilt is sometimes more upright, sometimes more on its side.
- The Earth's axis wobbles.
- These changes (called **Milankovitch Cycles**) could start or end an ice age.
- They affect the amount of sunlight the Earth receives.

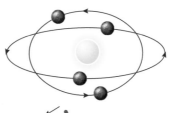

It takes 100 000 years for the Earth's orbit to change from being more circular, to an ellipse, and back again.

It takes 41 000 years for the Earth's axis to tilt, straighten up, and tilt back again.

It takes 26 000 years for the Earth's axis to wobble, straighten up, and wobble again.

Figure 1 *Orbital changes could cause climatic changes on long timescales*

Six Second Summary

- Ash and gas from large volcanic eruptions could alter the climate.
- Ash and dust from a large asteroid collision could block incoming sunlight.
- Changes in sunspots could cause warmer and cooler periods.
- Changes to the Earth's orbit could alter climate.

Over to you

Draw a mind map with 'Climate change in the past' in the centre. Add the four theories on this page and explain how each could change the climate.

Past climates – how do we know?

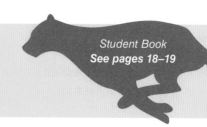

You need to know:

- how ice cores, tree rings and historical sources can tell us about past climates.

 Big Idea

Climate is the seasonal pattern of weather changes, brought by the world's winds, ocean currents and pressure systems.

How do we know about the past?

Evidence such as fossilised animals and trees, landforms and ice cores can be used to tell us about how climate has changed in the past (Figure 1).

Ice cores

Ice sheets contain layers of ice, with the youngest at the top. Taking an ice core can help to construct past climates.

- Air bubbles get trapped in the ice layers.
- The air bubbles contain carbon dioxide (CO_2).
- The amount of trapped CO_2 can tell us about past temperatures.
- Ice cores tell us there have been previous warm (inter-glacial) and cold (glacial) periods.
- We know about the last 2.6 million years (called the Quaternary period) from ice cores.

Tree rings

Each ring in a tree shows a year's growth.

- They tell us about the climate during a tree's lifetime.
- In warmer and wetter years, a tree grows more.
- Fossils of trees can go back thousands of years.

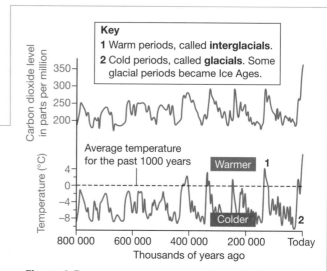

Figure 1 *Past temperatures and carbon dioxide levels*

Historical sources

These include historical drawings of the landscape, diaries or newspapers, or recorded dates of events, such as harvests.

- They are recent evidence of climate change.
- They may be inaccurate (because people see things in different ways)
- They suggest climate may change every few hundred years (Figure 2).

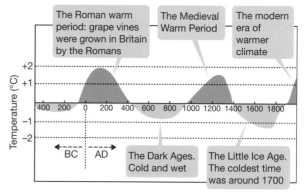

Figure 2 *The temperature has changed every few hundred years over the last 2000 years*

 Six Second Summary

- Climate is the seasonal pattern of weather changes.
- Ice cores, tree rings and historical sources tell us about climate change in the past.
- During the past 2000 years, we know that climate has changed every few hundred years.

Over to you

Write an answer for the question 'Explain how ice cores can tell us about past climates'. (4 marks)

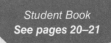

Student Book
See pages 20–21

You need to know:

- about the greenhouse effect
- how human activities produce greenhouse gases.

The greenhouse effect

The gases which make up the Earth's **atmosphere** are:

- nitrogen – important for plant growth
- carbon dioxide (CO_2) – a greenhouse gas taken in by plants
- oxygen – breathed in by animals
- water vapour – forms clouds.

The **greenhouse effect** (Figure 1):

- is completely natural
- is the way that **greenhouse gases** in the atmosphere trap heat from the sun
- warms the planet by 16°C, making it habitable.

Greenhouse gases include CO_2, methane, nitrous oxide and halocarbons.

- Extra greenhouse gases are produced by human activities.
- Burning fossil fuels produces CO_2 (the main source is from power stations producing electricity).
- CO_2 is also produced by industry (e.g. processing oil), transport (burning fossil fuels), farming (using fertiliser made from oil) and in heating homes.

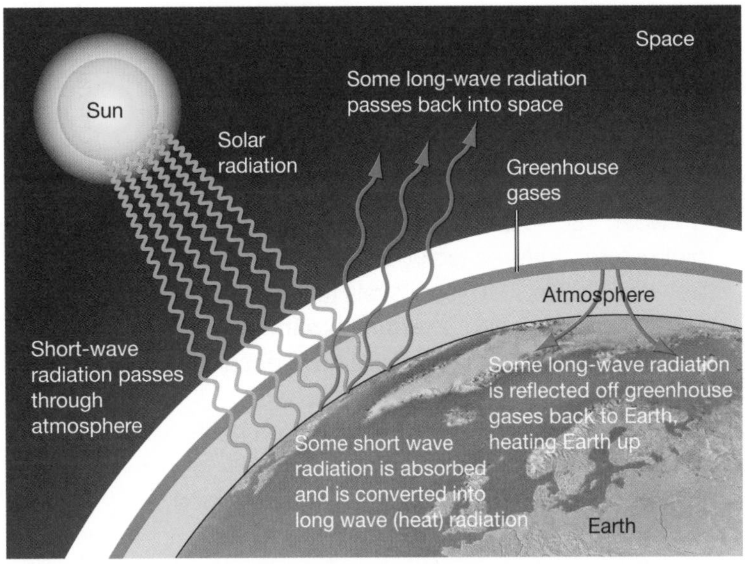

Greenhouse gases

Figure 2 shows global variations in CO_2 emissions. Scientists are concerned about the effect of CO_2 on climate. Their concerns include:

- reducing emissions in the developed world
- persuading developing countries to slow down their increase in emissions
- protecting vulnerable people from the impacts of climate change.

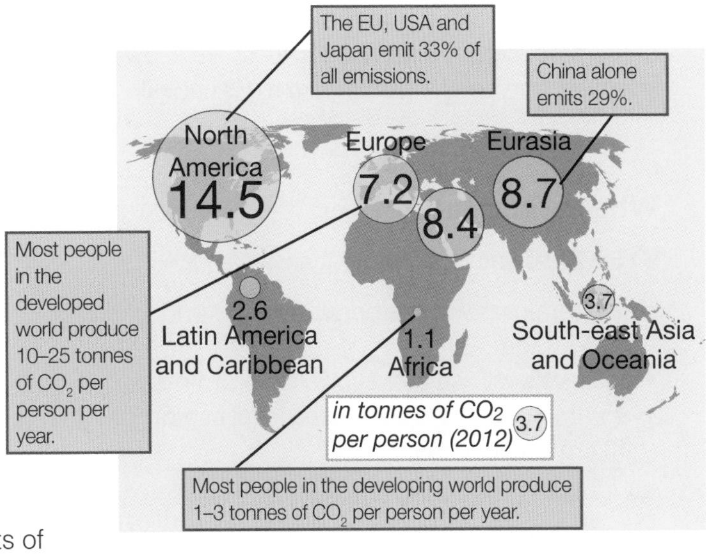

Figure 1 *The greenhouse effect - gases in the atmosphere act like the glass in a greenhouse, letting heat in, but preventing most of it from getting out*

The EU, USA and Japan emit 33% of all emissions.

China alone emits 29%.

North America **14.5**

Europe **7.2**

Eurasia **8.7**

8.4

Most people in the developed world produce 10–25 tonnes of CO_2 per person per year.

2.6
Latin America and Caribbean

1.1
Africa

3.7
South-east Asia and Oceania

in tonnes of CO_2 per person (2012) 3.7

Most people in the developing world produce 1–3 tonnes of CO_2 per person per year.

Figure 2 *CO_2 emissions vary around the world*

Six Second Summary

- The greenhouse effect is the way that gases in the atmosphere trap heat from the sun and warm the planet.
- Human activities are adding to the amount of carbon dioxide in the atmosphere.
- The developed world emits more carbon dioxide than the developing world.

Over to you

Make two lists, one with facts about the natural greenhouse effect, and the other to show which human activities add extra greenhouse gases to the atmosphere. Refer to pages 20–21 in the student book if you need to.

Student Book
See pages 22–23

You need to know:

- how atmospheric pollution has led to an enhanced greenhouse effect
- what might happen to the climate in the future.

A warming planet

Pollution of the atmosphere by greenhouse gases (Figure 1), is leading to an **enhanced greenhouse effect**. This causes climate change, often called global warming.

Global warming has been measured:

- Between 1880–2012, a 0.85°C rise in average global temperatures took place.
- Between 1870–2010, sea levels rose by 210 mm as the sea expanded (**thermal expansion**).
- The ten hottest years on record have occurred since 1998.
- Between 1979–2012, floating Arctic sea ice halved in size.
- 90% of the world's valley glaciers are shrinking.

What do climate experts think?

In 2014, the Intergovernmental Panel on Climate Change (a group of scientists) confirmed that warming is due to the increase in greenhouse gases produced by humans. But, there are a few scientists who believe global warming is natural.

What of the future?

By 2100, scientists estimate that:

- temperatures will rise between 1.1°C and 6.4°C (Figure 2)
- sea levels will rise between 30 cm and 1 metre
- a warming of 3.5°C and a sea level rise of 40 cm is a 'best guess'.

Predicting future global warming is difficult because we don't know how:

- population may grow
- fossil fuel use may change
- people's lifestyles may change.

Climate change could cause:

- more frequent floods and droughts
- stronger storms
- changes to farming
- people from low-lying areas becoming 'climate refugees'.

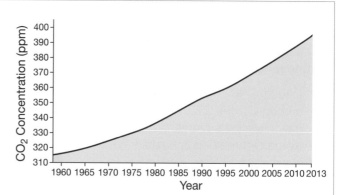

Figure 1 *Carbon dioxide concentrations in the atmosphere have been rising due to human activity. This graph shows recordings from Hawaii 1959–2013.*

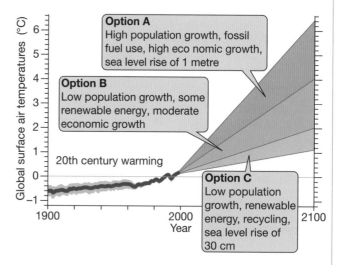

Figure 2 *Three possible increases in global temperatures by 2100, based on different human actions*

⏱ **Six Second Summary**

- Pollution of the atmosphere by greenhouse gases is causing the enhanced greenhouse effect.
- Global warming has increased more rapidly since 1980.
- Predicting future climate change is difficult, but a 'best guess' could see temperatures rise by 3.5°C.

✎ **Over to you**

Draw two mind maps, one on measuring climate change and the other on the impacts of climate change.

Student Book
See pages 24–25

You need to know:

- what tropical cyclones are, and what hazards they bring.

Cyclone alert!

On 19th February 2015, Australia's Northern Territory experienced winds of 140 km/h and gusts reaching 195 km/h. These were brought by a tropical cyclone named Lam.

- It was a very strong storm, defined as Category 4 by the Australian Bureau of Meteorology.
- It eventually died out on the 21st February.

What's in a name?

A tropical cyclone:

- is a rotating system of clouds and storms
- forms over tropical or subtropical waters
- has winds which can exceed 118 km/h
- is known as a hurricane, typhoon or cyclone and is measured differently depending on where in the world it originates (Figure 1).

Area of origin	Name	Measurement and classification
North Atlantic Ocean and Pacific coast of USA (area 1 on Figure 2)	Hurricane	**Saffir-Simpson Hurricane Scale.** It uses 5 categories to measure wind strength.
Indian and South Pacific Oceans (area 2 on Figure 2)	Cyclone	India has its own scale. Australia uses the Australian **Tropical Cyclone Intensity Scale**.
Western Pacific Ocean (area 3 on Figure 2)	Typhoon	Japan uses its own **Meteorological Agency's Scale**.

Figure 1 How tropical cyclones are named and measured

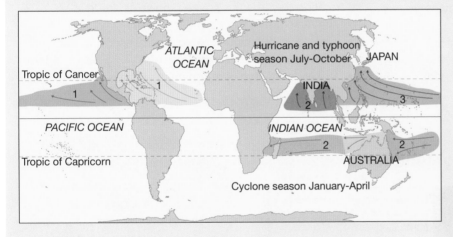

Figure 2 The location and pathways of tropical cyclones

What hazards do tropical cyclones bring?

Tropic cyclones bring a range of hazards.

- **Strong winds** – bring down trees, destroy power lines and even whole buildings.
- **Storm surges** – bring flooding (tides are higher because air pressure is so low).
- **Intense rainfall** – it is not unusual for 1000 mm of rain to fall.
- **Landslides** – intense rainfall leads to saturated, heavy ground, causing it to slump.

Six Second Summary

- Tropical cyclones are rotating systems of clouds and storms.
- They are known by different names and measured on different scales, depending on where they originate.
- They bring a range of hazards, such as strong winds, storm surges, intense rainfall and landslides.

Over to you

Describe **a)** three features of tropical cyclones, and **b)** three hazards they bring.

Student Book
See pages 26–27

You need to know:

- how a tropical cyclone forms.

Keeping track

Tropical cyclones:

- are relatively small (averaging 650 km, or 400 miles, across)
- form between the Tropics.

Once they form, their **track** (movement) is driven by global wind circulation and can generally be predicted using satellite photography, (see Section 1.4).

Big Idea

Tropical cyclones form under specific climatic conditions.

Where do tropical cyclones develop?

Tropical cyclones form in **source regions**. Their formation depends on three conditions occurring at the same time.

- A large, still, **warm ocean**, exceeding 26.5°C. This helps a warm body of air to develop.
- Strong winds, to draw warm air up rapidly from the ocean surface.
- A strong **Coriolis force**, created by the Earth's rotation.

Air pressure in tropical cyclones

Air pressure is the weight of air bearing down on the Earth's surface. Average air pressure is 1013 millibars.

- Tropical cyclones have much lower air pressure than air surrounding them because they are so warm.
- Big differences in air pressure creates strong winds rotating towards the centre (known as 'the eye').
- The moisture in the air causes huge towering cumulonimbus clouds, which spiral as the winds spin round towards the centre.
- The air in the eye of a tropical cyclone is calm and clear.

How does a tropical cyclone form?

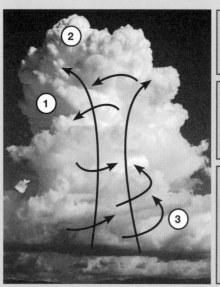

1 Warm air rises from the ocean, more air rushes in to replace it, then it too rises, drawn by the draught above.

2 Water vapour in the updraughts of air condense into **cumulonimbus clouds**. Heat energy from condensation powers the cyclone further.

3 Coriolis force causes the air to spiral around the centre of the cyclone. It rises and cools. Some descends as the **eye** of the storm. The cyclone enlarges as it is fed new heat and moisture. It loses energy and **decays** once it reaches a landmass.

Figure 1 *Stages in tropical cyclone formation*

Six Second Summary

- Tropical cyclones need warm oceans, strong winds and a strong Coriolis force to develop.
- The air pressure in a tropical cyclone is much lower than its surroundings.
- Tropical cyclones form with cumulonimbus clouds which spiral round a centre.

Over to you

- Name three conditions needed for a tropical cyclone to form.
- Explain **a)** why hurricanes have such fast winds, **b)** why they have such huge clouds.

The impacts of tropical cyclone Aila

You need to know:

- why Bangladesh is vulnerable to the impacts of tropical cyclones
- the impacts of Cyclone Aila on Bangladesh in 2009.

Student Book
See pages 28–29

Why is Bangladesh vulnerable to cyclones?

Bangladesh is hit by just 5% of the world's cyclones each year, but it suffers 85% of the world's deaths and damage caused by them. This is because:

- its people are poor; GDP per capita (PPP) was only $2100 in 2013

- 31% of its population lives below the poverty line
- the poorest are rural landless labourers, particularly women, forced to live on flood-prone land in poorly-built housing.

Cyclone Aila, May 2009

Tropical cyclones can have devastating impacts. This factfile explains what happened when Cyclone Aila hit Bangladesh on 25 May, 2009.

Factfile: Cyclone Aila, May 2009

Location
- Bangladesh.
- 60% the size of the UK. Population of 156 million.
- One of the world's poorest countries.

Vulnerability
- Lots of people live in areas most vulnerable to flooding (yellow area on map).
- Contains three of the world's largest rivers.
- Has an annual monsoon season.

Cyclone event
- Began as a tropical storm in the Bay of Bengal.
- High intensity of rain – 120 mm fell in a few hours.
- Strong winds reaching 360 km/h.
- Low pressure caused the sea level to rise, creating a huge **storm surge**.

Social and economic impacts
- 190 people killed.
- Storm surge destroyed several villages.
- Half of the flood protection embankments were washed away.
- 750 000 people made homeless.
- Crops destroyed in the delta.

- 3.5 million people affected.
- 59 000 animals killed affecting food supply and farming income.
- Sickness and typhoid spread from contaminated water.

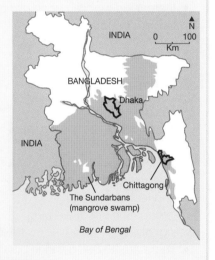

Environmental impacts
- Pressure on firewood resources as fuel from animal dung was lost.
- Freshwater contaminated by sewage.
- Sundarbans mangrove forest badly affected, 30 endangered tigers drowned.

Long term impacts
- By 2010, 200 000 were still living in temporary shacks.
- Poverty forced some people to migrate to slums in cities.

Six Second Summary

- Cyclones have devastating social, economic and environmental impacts.
- The poorest in vulnerable countries suffer the most.

Over to you

Create a table, with the headings social, economic and environmental impacts. Classify the impacts included in the factfile. Indicate if impacts fall in two categories.

You need to know:

- how effectively Bangladesh plans for, and responds to, tropical cyclones.

*Student Book
See pages 30–31*

How does Bangladesh protect its people?

Bangladesh attempts to protect the population from tropical cyclones using: forecasting, satellite technology, warning systems and evacuation strategies.

1 Weather forecasting

- Weather forecasts are issued through TV and radio.
- Outside Dhaka (the capital) few have TV or radios.
- Households with radios have lower death rates.
- Increasing mobile phone ownership helps to warn people.

2 Satellite technology

- Expensive weather satellite images from space are bought from US, China and Japanese satellites.
- Three radar stations transmit live weather updates to track cyclones.
- Bangladesh is spending US$150 million on its own space satellite.

3 Warning systems

- The government has developed an early warning system to evacuate coastal communities.
- Village meetings, posters and leaflets spread information about warning signals.
- There are 45 000 cyclone warning volunteers in vulnerable areas.

4 Evacuation strategies and storm surge defences

- Bangladesh has invested in evacuation procedures and safe refuges.
- 3500 brick or concrete cyclone shelters have been built – they can halve death rates.
- Embankments are built to protect against storm surges.

Measuring success

Bangladesh has successfully reduced the death tolls and damage from cyclones. But:

- warning systems are expensive
- illiteracy means that some cannot understand the warnings
- fear of losing property and previous false warnings stop some people from evacuating
- embankments cannot be built along the entire coastline and there are not enough cyclone shelters.

 Six Second Summary

- Bangladesh prepares for cyclones using forecasting, satellite technology, warning systems and evacuation strategies.
- Bangladesh has been successful, but more could be done.

Over to you

Create a mnemonic to remember Bangladesh's ways of planning and preparing for cyclones.

Student Book
See pages 32–33

You need to know:

- how effectively the USA plans for, and responds, to tropical cyclones.

Managing hurricane risks in the USA

The USA experiences **hurricanes** (another name for tropical cyclones) between July and October along its Atlantic coast. Florida has a 22% chance of experiencing a hurricane each year.

The USA protects its people in three main ways.

2 Risk and evacuation

Warning and evacuation systems help plan evacuations.

- In Florida, towns and cities are classified into **risk zones** (Figure 1) for high winds or storm surges.
- Only people who need to leave are evacuated.
- This helps by not overwhelming emergency services.

3 Storm surges and defences

Hurricane Katrina in 2005 was the worst hurricane in US history.

- 1833 people died.
- Damage cost US$108 billion.
- Artificially-built **river levées** (banks) collapsed, flooding 80% of New Orleans with a 4 m high storm surge.
- The levées had been poorly maintained.
- 80% of the population of New Orleans was evacuated.
- The poor and elderly were left behind.

There are now 'softer' and cheaper defences available to protect against storm surges, such as beach nourishment and wetland creation which absorb water and wave energy.

1 Forecasting, warning and satellite technology

The National Hurricane Center in Miami issues forecasts, warnings and educates people about tropical cyclones.

- The USA has over 20 weather satellites.
- Weather forecasts are issued frequently on TV and radio.
- Almost everyone has access to the media and a mobile phone.
- Some satellites are ageing and failed to properly track Hurricane Sandy in 2012.

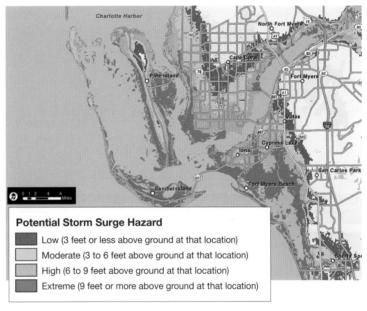

Potential Storm Surge Hazard

- Low (3 feet or less above ground at that location)
- Moderate (3 to 6 feet above ground at that location)
- High (6 to 9 feet above ground at that location)
- Extreme (9 feet or more above ground at that location)

Figure 1 *Towns and cities in Florida are classified into* **risk zones**

 **Six Second Summary**

- The USA prepares for cyclones through forecasting, satellite technology, warnings, evacuation systems and storm surge defences.
- Some satellites and defences need updating and maintaining.
- Hurricane Katrina was the worst cyclone to hit the USA.

 **Over to you**

List two differences and two similarities between the USA and Bangladesh's planning and preparation for tropical cyclones.

Student Book
See pages 34–35

Journey to the centre

No one has ever seen inside the Earth (Figure 1), we would have to drill 6365 km to reach its centre. What we know about the Earth's interior comes from:

• direct evidence from the Earth's surface
• indirect evidence from earthquakes e.g. studying the speed and movement of seismic (earthquake) waves as they travel through the Earth can tell us about the physical state of the different layers
• indirect evidence from material from space.

Figure 2 shows the properties of the Earth's layers.

Big Idea

The structure of the Earth is key to plate tectonics.

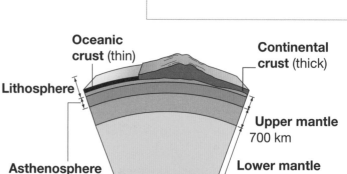

Figure 1 *The structure of the Earth is layered, like an onion*

	Layer	Density (grams/cm³)	Physical state	Composition	Temp (°C)	Features
Lithosphere	Continental crust	2.7	Solid	Granite	900	• The uppermost layer of the Earth's surface. • Split into tectonic plates. • Moves on the asthenosphere (the top layer of the mantle). • **Continental crust** forms the land and is thicker. • **Oceanic crust** is found under the oceans and is thinner.
	Oceanic crust	3.3	Solid	Basalt	900	
Mantle	Asthenosphere	3.4–4.4	Partially molten	Peridotite	900–1600	• The 'lubricating' layer under the lithosphere.
	Lower mantle	4.4–5.6	Solid		1600–4000	• The largest of the Earth's layers.
Core	Outer core	9.9–12.2	Liquid	Iron and nickel	4000–5000	• Evidence from meteorites (fragments of rock and metal from a shattered planet that fell from space) tell us about its composition. • The inner core is solid and 2900 km below the surface.

Figure 2 *The properties of the Earth's layers*

 Six Second Summary

• The Earth is divided into layers.
• Each layer has different compositions and properties.
• Evidence for the Earth's structure comes from its surface, earthquakes and from space.

 Over to you

• Create a mnemonic to help you remember the Earth's layers.
• From memory, sketch the structure of the Earth and label each layer with one fact about its properties.

*Student Book
See pages 36–37*

You need to know:

- how the core's internal heat source generates convection, and moves plates.

Hot rocks

We know that the inside of the Earth is hot because of:

- molten lava from active volcanoes
- hot springs and geysers.

Geothermal ('earth-heat') is produced by the **radioactive decay** of elements in the core and mantle. The core can reach over 5000°C.

Figure 1 shows how the movement of tectonic plates occurs.

- Heat rises from the core and creates **convection currents** in the outer core and mantle.
- It moves very large 'slabs' of crust.

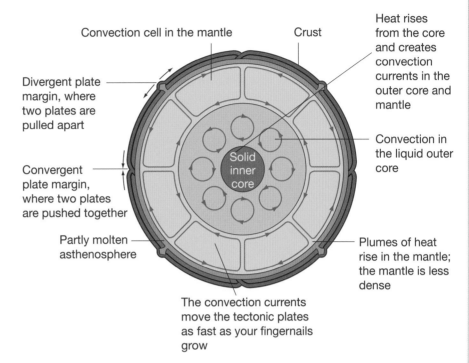

Convection cell in the mantle — Crust — Heat rises from the core and creates convection currents in the outer core and mantle

Divergent plate margin, where two plates are pulled apart

Convection in the liquid outer core

Convergent plate margin, where two plates are pushed together

Solid inner core

Partly molten asthenosphere

Plumes of heat rise in the mantle; the mantle is less dense

The convection currents move the tectonic plates as fast as your fingernails grow

Figure 1 *How plate tectonics is driven by convection currents*

Plumes

Plumes make up part of the convection cell. These plumes:

- are concentrated zones of heat which move towards the surface
- bring **magma** (molten rock) to the surface
- erupt as **lava** in a volcano if they break the surface
- form **divergent plate boundaries** if they rise like long sheets of heat (Figure 1)
- form **hot spots** on the surface where they rise like columns of heat.

Magnetic field

A huge invisible magnetic field (the magnetosphere) surrounds the Earth. This magnetic field:

- can sometimes be seen as the **aurora borealis** (northern lights), formed when radiation from space hits the magnetosphere
- protects the Earth from harmful radiation
- is created by the outer core – as liquid iron flows, it works like an electrical dynamo.

Six Second Summary

- The Earth is heated by radioactive decay in the core and mantle.
- Convection currents created by the heat move tectonic plates.
- Rising heat creates plumes which bring magma to the surface.
- The Earth's magnetic field is created by the outer core and protects the Earth from harmful radiation.

Over to you

Answer the question 'Explain how the core's internal heat moves tectonic plates on the Earth's surface'. (4 marks)

Student Book
See pages 38–39

You need to know:

- how tectonic plates have moved in the past
- the three types of plate boundary.

Pangea, the supercontinent

Scientists believe that the Earth's present continents were once all joined together forming a supercontinent called Pangea.

- Identical rocks and fossils from 250 million years ago have been found in West Africa and South America, proving they were joined.
- Pangea began to separate about 200 million years ago.
- Plate tectonics have moved the continents to their present positions.

Crust: old and new

Continental crust is 3–4 billion years old. The oldest oceanic crust is only 180 million years old. This is because:

- new oceanic crust forms at divergent plate boundaries (Figure 2)
- old oceanic crust is destroyed by **subduction** at convergent plate boundaries
- continental crust was formed billions of years ago and has not formed since.

Moving plates

The Earth's lithosphere is split into 15 large and 20 smaller **tectonic plates**. They move slowly on the asthenosphere. The place where they meet is a **plate boundary** (Figure 1), where most earthquakes and volcanoes are found.

- Divergent plate boundaries occur where two plates move apart.
- Convergent plate boundaries occur when two plates collide.
- Conservative plate boundaries occur when two plates slide past each other.

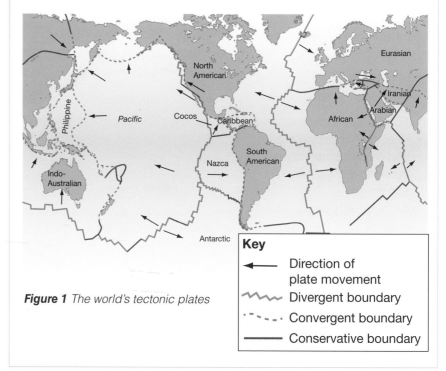

Figure 1 *The world's tectonic plates*

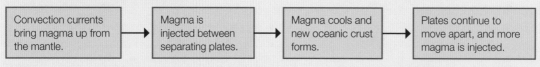

Figure 2 *How new oceanic crust forms at divergent plate boundaries*

Six Second Summary

- The continents were once joined together, but plate tectonics have moved them apart.
- Plate boundaries occur where two plates meet.
- New oceanic crust is constantly formed at divergent plate boundaries.

Over to you

Explain the difference between convergent, divergent, and conservative plate boundaries.

Boundary hazards

You need to know:

- which hazards occur at different plate boundaries.

*Student Book
See pages 40–41*

Tectonic hazards

Earthquakes and volcanoes are **tectonic hazards.** They occur at plate boundaries. Figure 1 shows the hazards at each type of plate boundary. Figure 2 compares divergent and convergent plate boundaries.

Plate boundary	Example	Earthquakes	Volcanoes
Conservative – plates slide past each other	San Andreas fault in California, USA	• **Friction** between the plates causes earthquakes. • They are shallow, destructive, up to magnitude 8.5. • Small tremors almost daily.	• None.
Divergent – plates move apart and magma rises to fill the gap	Iceland sits on the mid-Atlantic ridge	• Small earthquakes up to magnitude 6.0. • Caused by **friction** as the plates tear apart.	• Hot and runny magma made of **basalt.** • **Lava flows** and shallow sided volcanoes. • Not very explosive or dangerous.
Convergent – plates push together and the oceanic plate is **subducted**	Andes mountains in Peru and Chile	• Pressure builds up if the sinking oceanic plate sticks to the continental plate. • Earthquakes can be violent and devastating, up to magnitude 9.5. • Tsunami can form.	• Melting oceanic plate creates magma called **andesite.** • Andesite is less dense so rises in **plumes.** • Water erupts as steam, so the volcanoes are explosive. • Volcanic ash and 'bombs' are blasted up as **pyroclasts.** • Steep-sided, cone shaped volcanoes.
Collision zone – a convergent boundary when two continental plates collide, folding sediments into mountains	Himalayas	• Destructive earthquakes up to magnitude 9, e.g. the 2015 Nepal earthquake. • Can trigger landslides.	• Very rare.

Figure 1 *Features of plate boundaries*

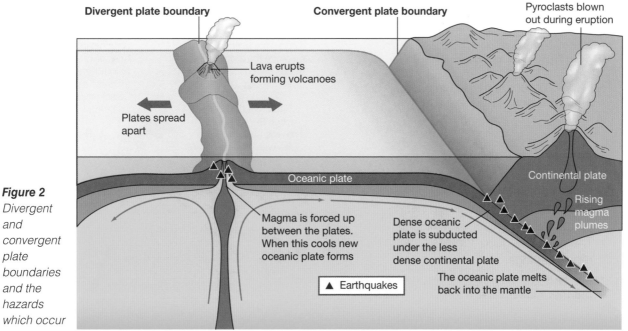

Figure 2 *Divergent and convergent plate boundaries and the hazards which occur*

- Earthquakes occur at all types of plate boundary, but vary in their magnitude.
- Volcanoes occur mainly at divergent and convergent plate boundaries.

Over to you

In 10 seconds for each, explain what occurs at the following plate boundaries:

- convergent
- divergent
- conservative
- collision.

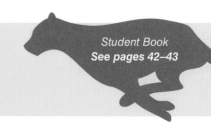

Student Book
See pages 42–43

You need to know:

- the impacts of volcanoes on property and people in developed countries.

Destructive power

The more explosive a volcano, the more destruction it causes.

- Destructive power is measured on the **Volcanic Explosivity Index (VEI)** on a scale from 1 to 8.

Volcanoes produce:

- **primary effects** – caused instantly by the eruption e.g lava, lahars
- **secondary effects** – occur in the hours, days and weeks afterwards e.g. disease.

Sakurajima, Japan

Japan is on a convergent plate boundary, so there are active volcanoes. Sakurajima has erupted since the 1950s, sometimes 200 times a year.

The volcano

- Sakurajima is a stratovolcano, formed of layers of ash and lava.
- 7000 people live at its base.
- The nearest city, Kagoshima, has a population of 650 000.
- Hot springs and lava flows are popular with tourists.
- Surrounding soils are very fertile.

Effects of eruptions

- Volcanic bombs are hurled over 3 km.
- There are 2 km-long pyroclastic flows.
- 30 km³ of ash erupts each year.
- Poisonous gases bring acid rain, which kills plants.
- Ash and lava bury buildings and farmland.

Planning and preparation

- Eruptions can be **predicted** so people can be **evacuated.**
- Sakurajima is monitored (Figure 1) by seismologists.
- Japan is a developed country and can afford to spend money on protection, such as lava bomb shelters.
- People rarely die from eruptions.
- People have insurance on their homes.
- The government contributes funding to repair damage.
- Damage occurs to property (economic), but there is less harm to people (social).

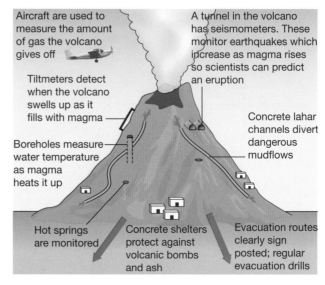

Aircraft are used to measure the amount of gas the volcano gives off

Tiltmeters detect when the volcano swells up as it fills with magma

Boreholes measure water temperature as magma heats it up

Hot springs are monitored

Concrete shelters protect against volcanic bombs and ash

A tunnel in the volcano has seismometers. These monitor earthquakes which increase as magma rises so scientists can predict an eruption

Concrete lahar channels divert dangerous mudflows

Evacuation routes clearly sign posted; regular evacuation drills

Figure 1 *Sakurajima - monitoring and evacuation*

 **Six Second Summary**

- Volcanic eruptions have primary and secondary effects.
- Sakurajima erupts regularly, but is well monitored.
- Volcanic hazards in developed countries have high economic costs, but relatively low social costs.

Over to you

Draw a spider diagram to explain
a) on one 'leg', the risks of living near Sakurajima, and **b)** on a second 'leg', why people continue to live near the volcano.

Student Book
See pages 44–45

You need to know:

- the impacts of volcanoes on property and people in developing countries.

At risk

People in developing countries are at greater risk from tectonic hazards than those in developed countries because:

- poorly-constructed (cheaper) housing is more likely to be built in risky locations (which are more affordable)

- people are much less likely to afford insurance
- governments do not have the revenue to provide **aid**
- communications are poor, making it harder to evacuate, with limited access for emergency services.

Mount Nyiragongo, Democratic Republic of the Congo (DRC)

Mount Nyiragongo is a large stratovolcano in DRC. It erupted in January 2002.

The eruption

- A fast-flowing river of basalt lava poured into the city of Goma.
- Poisonous gases were emitted.
- Water supplies were disrupted.

Impacts of the eruption

- 100 people died.
- 12 500 homes were destroyed.
- Over 120 000 people were made homeless.
- 400 000 people were evacuated to overcrowded refugee camps.
- There was concern about the spread of diseases such as cholera.
- Acid rain affected farmland.
- Most people could not afford to rebuild their homes.

Relief effort:

- The United Nations and Oxfam began a **relief effort.**
- Families each received 26 kg of food rations.
- Governments around the world gave US$35 million.
- Emergency measles vaccinations were given.
- By June, some roads had been cleared and water supplies were repaired.

Future threats

- Mount Nyiragongo could erupt at any time.
- Gases, such as suphur dioxide, rise into nearby Lake Kivu and become trapped. Earthquakes release them, and people can suffocate.
- Threats are different to those from Sakurajima (Figure 1).

	Mt Sakurajima, Japan	Mt Nyiragongo, DRC
Volcano type	Steep-sided, high, stratovolcano	High but less steep, stratovolcano
Magma type	Andesite, high viscosity	Basalt, low viscosity
Explosivity	VEI 4–5	VEI 1
Hazards	Lava flows, volcanic bombs, pyroclastic flows, constant eruptions	Lava flows, gas emissions

Figure 1 *Comparing Mount Sakurajima and Mount Nyiragongo*

Six Second Summary

- Volcanic eruptions usually have more devastating consequences in the developing (rather than the developed) world.
- Poverty means people in the developing world are more vulnerable, and less able to cope with eruptions.

Over to you

List two social, and two economic, impacts of the 2002 Mount Nyiragongo eruption.

Earthquake!

- about earthquake hazards and their impacts
- how long-term planning can manage the hazards.

Student Book
See pages 46–47

Why is the ground shaking?

Earthquakes can very rarely be predicted, and are sometimes catastrophic.

- They result from a sudden release of energy.
- The **magnitude** of an earthquake is how much the ground shakes.
- A **seismometer** measures magnitude on the **Richter scale**.
- The scale is logarithmic – a 6.0 quake is 10 times more powerful than a 5.0, and so on.
- Energy travels out from the **focus** (origin) in waves.
- The **epicentre** is directly above the focus, on the Earth's surface.
- The shallower the focus, the more destructive the earthquake.

Tsunami

Earthquakes beneath the sea bed can generate a tsunami (Figure 1). Warning systems can detect them, but only if they are far from a coast. They can destroy homes and infrastructure.

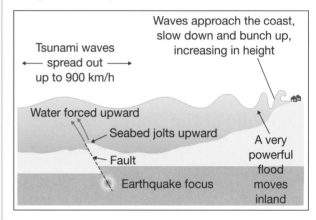

Figure 1 How a tsunami forms

Comparing Haiti and Japan

Major earthquakes affected Haiti and Japan within a year of each other, but with different causes and effects.

	Sendai, Japan 2011	Port-au-Prince, Haiti 2010
Earthquake	• Magnitude - 9.0 • Focus – 30 km deep • Convergent plate boundary • Epicentre – 70 km from the coast	• Magnitude - 7.0 • Focus – 13 km deep • Conservative plate boundary • Epicentre – 25 km from the capital city
Primary effects	• 1 dam collapsed • 2 nuclear power stations fractured • An oil refinery set on fire • Motorway damaged and airport closed • US$235 billion of damage	• Deaths - 316 000 • Injured - 300 000 • Homeless - 1 million • The port, communications and roads were damaged beyond repair
Secondary effects	Caused by the tsunami: • Deaths - 15 900 • Missing – 2600 • Injured - 6150 • Homeless - 350 000 • Two nuclear reactors went into meltdown • Businesses disrupted • Disrupted schooling, unemployment and stress lasted for years	• Cholera outbreak killed over 8000 people • Aid struggled to get in • 1 in 5 jobs were lost from clothing factories • By 2015 most people displaced had been rehoused

Figure 2 Comparing Haiti and Japan

Long-term planning

There is a 70% **probability** that a magnitude 7.2 earthquake will hit Tokyo in the next 30 years. But, this cannot be **predicted**. Japan is a developed country, so invests in long-term planning, earthquake drills, emergency kits, building design and tsunami walls.

Six Second Summary

- The magnitude of an earthquake is measured on the Richter scale.
- Tsunami occur when an earthquake happens beneath the sea bed.
- Developed countries can cope with, and plan for, earthquakes more effectively than developing countries.

Over to you

Create two mind maps to show the effects of the Haiti and Japanese earthquakes.

Student Book
See pages 48–49

You need to know:

- the impacts of earthquakes on developing countries
- how people respond to earthquakes.

Death and destruction

Destructive earthquakes in the developing world happen fairly frequently. They tend to have much higher death tolls in comparison to volcanic eruptions. Nepal suffered from two devastating earthquakes in 2015.

 Big Idea

Tectonic hazards have a greater impact in developing than developed countries.

Factfile: Nepal earthquakes, 2015

Location
- One of Asia's poorest countries.
- Rural and isolated.
- Surrounded by the Himalayas.

Event
- Earthquake magnitude 8.1 on 25 April.
- Earthquake magnitude 7.3 on 12 May.
- Aftershocks reaching magnitude 6.7.
- Collision between the Indian and Eurasian plate.
- Epicentre was along the main frontal thrust of plate boundary, 6 km deep.

Social impacts
- 9107 people died, including 329 in a landslide and 19 on Mount Everest.
- 23 000 injured, 6000 still being treated a month later.
- Several hundred thousand made homeless.
- Drinking water and sanitation systems destroyed.

Economic impacts
- International airport in Kathmandu, the capital, was closed, preventing aid deliveries.
- Rebuilding costs of US$7 billion.
- Crop planting season was missed and rural families lost incomes.

Environmental impacts
- 1000 ancient temples destroyed.

International responses
- It was estimated that Nepal could never repay loans, so relied on aid gifts.
- Half of the money was given by the Asia Development Bank.
- The UN provided blankets, tents and water.
- Indian troops searched for people and cleared rubble.
- The UK government donated US$51 million.
- The UK public donated US$80 million.
- The US and China provided helicopters to access mountainous areas.
- Several organisations helped with rebuilding cheap houses to withstand earthquakes (Figure 1).

 Six Second Summary

- The impacts of earthquakes in the developing world are more devastating than those in developed countries.
- Developing countries rely on international responses and aid donations.

Over to you

Identify the meaning of the following numbers in relation to the Nepal earthquake:

- 6000
- 6
- 8.1
- 80 million

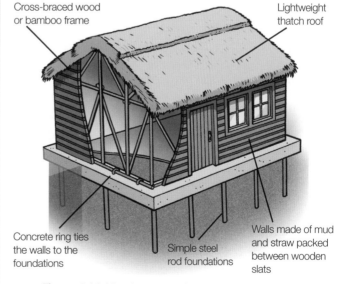

Figure 1 *Making houses safer*

Labels: Cross-braced wood or bamboo frame · Lightweight thatch roof · Concrete ring ties the walls to the foundations · Simple steel rod foundations · Walls made of mud and straw packed between wooden slats

Topic 2
Development dynamics

Your exam

- Topic 2 Development dynamics makes up Section B in Paper 1, Global geographical issues.

- Paper 1 is a 90-minute written exam and makes up 37.5% of your final grade. The whole paper carries 94 marks (including 4 marks for SPaG) – questions on Topic 2 will carry 30 marks.

- You have to answer all questions in Paper 1. Section A contains questions on Hazardous Earth (pages 12–33), and Section C on Challenges of an urbanising world (pages 50–63).

Your revision checklist

Tick these boxes to build a record of your revision

Spec Key Idea	Detailed content that you should know	1	2	3
2.1 There are different ways of defining and measuring development	• Defining and measuring development			
	• How countries at different levels of development differ in population data			
2.2 Global inequality in development and theories of how it can be reduced	• Causes and consequences of global inequalities			
	• Rostow's modernisation theory and Frank's dependency theory			
2.3 Approaches to development vary in type and success	• Characteristics of top-down and bottom-up strategies; why some countries benefit more than others			
	• Advantages and disadvantages of different approaches to development			
2.4 Development of the emerging country is influenced by its location and context in the world	• Site, situation and connectivity of the country			
	• Broad political, social, cultural and environmental background of the country			
2.5 Globalisation causes rapid economic change in the emerging country	• Key economic trends since 1990			
	• The role of globalisation and government policy in the development of the country			
2.6 Rapid economic growth results in significant positive and negative impacts on people and environment	• How rapid economic change has contributed to demographic change, urbanisation, and unequal wealth patterns			
	• Impacts of economic development and globalisation on different groups			
2.7 Rapid economic development has changed the international role of the emerging country	• How rapid economic development has changed its geopolitical influence			
	• Costs and benefits of changing relations, and the role of foreign investment			

Student Book
See pages 52–53

Measuring development

Malawi is one of the world's poorest countries. The United Nations (UN) uses economic and social **development indicators** to work this out (Figure 1).

	Malawi	Brazil	UK
GDP per capita (US$) measured in PPP	900	15 200	37 300
Access to safe drinking water %	83	97.5	100
Literacy rate %	61	91.3	99
HDI	0.41	0.74	0.89

Figure 1 *Socio-economic development indicators for Malawi in 2014, compared to a middle-income country (Brazil) and high-income country (the UK)*

Economic development indicators include:

- **GDP (Gross Domestic Product)** – The total value of goods and services produced by a country in a year (in US$).
  - Dividing GDP by population gives **GDP per person** (or *per capita*).
  - GDP is measured in PPP (**Purchasing Power Parity**) which shows what money will buy in each country.

- **Measures of inequality** – These show how equally wealth is shared among the population.

Social indicators include:

- **Access to safe drinking water** – The percentage of the population with a piped water supply within 1 km.
- **Literacy rate** – The percentage of the population, aged over 15, who can read and write.

The HDI

The **HDI** (Human Development Index) consists of one figure per country, between 0 and 1 (the higher the better). It is calculated using an average of four indicators:

- life expectancy
- literacy
- average length of schooling
- GDP per capita (using PPP$).

The poorest countries in terms of GDP often have the lowest HDI because they don't have money for health care and education.

Corruption and development

Corruption can get in the way of economic development if aid or investment is used to bribe officials, or purchase weapons. The Corruption Perceptions Index (Figure 2) uses a scale from 10 (honest) to 0 (very corrupt) to rank countries on how corrupt their governments are believed to be.

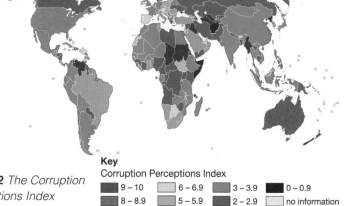

Figure 2 *The Corruption Perceptions Index*

Key
Corruption Perceptions Index

9 – 10	6 – 6.9	3 – 3.9	0 – 0.9
8 – 8.9	5 – 5.9	2 – 2.9	no information
7 – 7.9	4 – 4.9	1 – 1.9	

Six Second Summary

- Economic, social and political indicators help to measure development.
- HDI is calculated using an average of four indicators.
- Corruption can prevent economic development.

Over to you

From memory, write definitions of:

- **three** economic indicators
- **two** social indicators
- **one** political indicator.

Student Book
See pages 54–55

You need to know:

- how populations change as countries develop, with examples from Malawi.

Malawi's population

Malawi has the world's most **youthful population** (Figure 1).

- Its **demographic data** (Figure 2) are typical of many of the world's poorest countries.

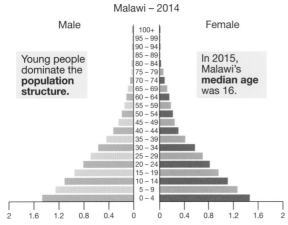

Malawi – 2014

Male | Female

Young people dominate the **population structure.**

In 2015, Malawi's **median age** was 16.

Population (in millions)　Age Group　Population (in millions)

Figure 1 *Malawi's population structure*

Data	Malawi	Brazil	UK
Population			
Birth rate (per 1000 people) (2014)	41.8	14.7	12.2
Death rate (per 1000 people) (2014)	8.7	6.5	9.3
Fertility rate (2014)	5.7	1.8	1.9
Dependency ratio	93.3	45.8	54.8
Health			
Life expectancy (years)	60	73.3	80.4
Infant mortality per 1000 live births	48	19.2	4.4
Maternal mortality per 100 000 births (2010)	460	69	12
Education			
Literacy rate %	74.8	91 (2012)	99

Figure 2 *Comparing demographic data in 2013 for Malawi, Brazil and the UK*

Population change

As countries develop, GDP per capita increases. This means more money to spend on health, education and piped water supplies. Figure 2 shows that as development increases:

- birth and death rates, dependency ratios, fertility rates, infant mortality rates and maternal mortality rates all fall
- life expectancy and literacy rates increase.

As birth rates fall and life expectancy increases, populations age.

Women's health and education

Malawi's infant and maternal mortality rates are among the world's worst, though they have halved since 2000.

- Skilled health workers attend only 20% of births.
- Babies of unhealthy mothers are more likely to die in their first five years.

Most secondary schools charge fees which subsistence families cannot afford.

- Few girls attend secondary school beyond age 13.
- In rural areas, many marry by 14, and have their first child soon after. Malawi's high fertility rate is caused by poverty.

Professional women in Malawi have fertility and infant mortality rates more like those of high-income countries because they are better educated and have children later.

Six Second Summary

- As wealth increases development indicators improve.
- Demographic indicators include birth rate and fertility rate.
- Malawi's high fertility rate is due to poverty.

Over to you

Cover everything on this page apart from Figure 1. Use the population pyramid to explain the population structure of Malawi.

Global inequality

Student Book
See pages 56–57

You need to know:

- about global inequality and whether it has changed in recent years.

Mind the gap

In 1980, the Brandt Report identified:

1 Wealthy countries in the '**global north**' (blue on Figure 1). They were **High Income Countries** (HICs).
2 Poorer countries in the '**global south**' (red on Figure 1). They were **Low Income Countries** (LICs).

These inequalities became known as the 'North-South Divide' or the 'Development Gap'.

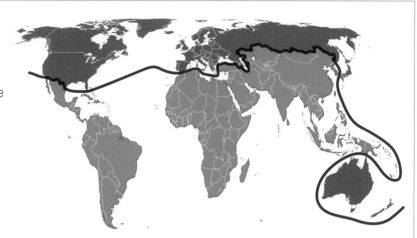

Figure 1 *The 'North-South Divide', from the Brandt Report*

Global shares of wealth

Every year, the World Bank ranks countries by total GDP. They are then grouped into **quintiles** (or fifths).

The results show that:

- the top quintile – the richest fifth of the world's countries – owns 83% of the world's wealth
- the bottom two quintiles – the poorest 40% - own just 3%.

A new world order

The world's ten largest economies in 2015 have changed since 1980. For example, China has moved from 9th to 2nd place, and Brazil has moved from 12th to 7th. But two things have stayed the same:

- the USA's GDP still far exceeds the rest
- the world's poorest countries are still in sub-Saharan Africa.

What's changed?

Since the Brandt Report was published, many **emerging economies** have become wealthier (Figure 2).

Decade of rapid growth	Area of rapid development	Main reason for development	Name given to these countries now
1980s	Latin America – e.g. Brazil	Large reserves of raw materials encouraged investment.	**Middle Income Countries** (MICs)
1990s	South-east Asia – e.g. Hong Kong and Singapore	Relocation of manufacturing overseas by US and European TNCs.	Growth was so aggressive that these countries became known as the 'Asian Tigers'. Most are now classed as **Newly Industrialising Countries** (NICs).
2000s and 2010s	China and India	Recent investment by US and European TNCs. Large domestic populations.	**Recently Industrialising Countries** (RICs). Together with Brazil and Russia, these countries are also known as the BRICs.

Figure 2 *Rapid growth in emerging economies*

Six Second Summary

- The Brandt Report divided the world into HICs and LICs.
- There is a 'development gap' between the world's richest and poorest countries.
- Since the 1980s, MICs, NICs and RICs have emerged.

Over to you

On a blank world map, draw the line of the 'North-South Divide'. Shade, in different colours, the MICs, NICs and RICs.

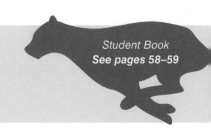

You need to know:

* how, and why, Malawi faces physical and environmental barriers to its development.

Student Book
See pages 58–59

Malawi's uphill task

Malawi faces increasing food prices and fuel shortages.

* Without investment, GDP remains low, and there is little tax to spend on education or health care. It's a **vicious cycle** (Figure 1).
* Malawi faces an uphill task in joining a globalised world.

What's holding Malawi's development back?

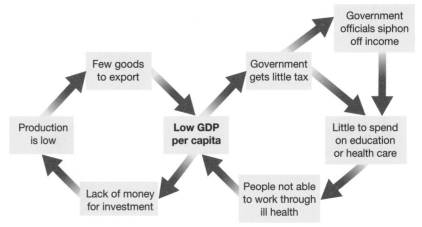

Figure 1 *The economic cycle of poverty*

1 It's landlocked

* Malawi has no coastline, so has no port from which to export or import goods.
* The nearest port is Nacala in Mozambique - and involves a slow, 800 km, single track railway.
* Nearly all Malawi's exports go this way. The trains return with imports. But it's a slow, expensive process.

2 Rural isolation

* 85% of Malawi's population is rural and isolated.
* Roads are mostly dirt, so it takes several hours to travel to markets in the rainy season.
* Mobile phone ownership is growing rapidly. It should make rural areas less isolated but rural coverage is poor.

3 Living with a changing climate

Oxfam published a report showing that climate change would cause:

* water shortages as temperatures rise (increasing evaporation)
* food shortages caused by variable rainfall.

Rainfall in Malawi has been much lower since 2000.

* The rainy season has been shorter; crop yields have fallen.
* When rains arrive, they are intense– damaging and reducing harvests.

4 Increased pollution

Rapid economic growth in Malawi since 2010 has led to urban growth.

* Squatter settlements have grown rapidly, with no sanitation or waste management.
* Rivers have become contaminated, causing risk to human health.
* Dust, industrial smoke, and car exhaust fumes reduce air quality.

 Six Second Summary

Four factors hold back Malawi's development:

* being landlocked affects trade
* rural communities are isolated
* climate change reduces crop yields
* pollution causes risks to human health.

 Over to you

Write each of the numbered titles above on revision cards. For each title, summarise in one sentence why that factor is causing Malawi's development to be held back.

You need to know:

- how Malawi faces economic and political barriers to its development.

The economic barriers ahead

Malawi can only develop by increasing trade. Three problems stand in the way.

1 Terms of trade

- The value of Malawi's exports is less than its imports so it earns less than it spends.
- Its **terms of trade** are stacked against it (Figure 1).
- One of the reasons is that Malawi exports largely raw materials **(primary products)**.
- It traditionally sold these to developed countries, and in return bought manufactured goods. This was typical of trade in the 1980s.
- Now, trade between developing countries and the emerging economies of India and China is also important.

Exports	Imports	Debt total
Value: $1.3 billion (2014)	**Value:** $2.5 billion (2014)	$1.729 billion (2014) $1.487 billion (2013)
Goods sold	**Goods bought**	
Tobacco (53%), tea, sugar, cotton, coffee, peanuts, and wood products	Food, petroleum products, consumer goods, transport equipment	
Sold mainly to	**Bought from**	
Developed and emerging countries: *Canada 12.4%, South Africa 6.7%, USA 6.4%, Russia 6.3%, Germany 6.1%, South Korea 4.3%*	Developed and emerging countries: *South Africa 25.1%, China 12.7%, India 12%*	
Developing countries: *Zimbabwe 9.4%, Zambia 6.1%*	Developing countries: *Zambia 11.7%, Tanzania 4.8%*	

Figure 1 *Malawi's terms of trade, and debt*

2 Colonisation and cash crops

Over 80% of Malawi's population works in farming.

- Malawi depends on **cash crops** (known as **commodities**) for exports.
- Global prices vary. Farmers never know what price they will get.

In the nineteenth century, the British **colonised** (ruled) Malawi.

- Plantations that they developed are still owned by UK owners – some are large TNCs (e.g. Unilever, producing PG Tips tea).
- Profits go to TNCs in developed countries – an example of **neo-colonialism**.
- In the UK tea is sold at around £8 per kg – 800 times the price paid to farmers in Malawi.

3 Global trade and international relations

The **World Trade Organisation** (WTO) tries to help developing countries trade with wealthier countries. It aims to ensure goods will be free of duties, or **tariffs**. It doesn't always work.

- Malawi exports raw coffee beans instead of roasting them which would get a higher price. This is because the EU and USA charge tariffs of 7.5% on imported roasted beans, but nothing on raw beans.
- It's cheaper for coffee companies (e.g. Starbucks) to roast beans rather than buy ready-roasted from Malawi.

 Six Second Summary

- Malawi's terms of trade are stacked against it.
- Malawi depends on cash crops, and prices constantly change.
- Tariffs mean Malawi continues to export raw coffee beans.

 Over to you

From memory, write definitions for each of the terms in bold on this page.

Student Book
See pages 62–63

You need to know:

- about two theories regarding how countries develop over time.

Why are some countries poor?

There are different theories about the causes of poverty.

- Rostow's theory sees a path from poverty to wealth that countries have to follow.
- Frank believes that a country's poverty is due to its past relationships with other countries.

Rostow's theory

In 1960, Walt Rostow published his theory.

- He believed that countries should pass through five stages of development (Figure 1).
- It became known as modernisation theory.

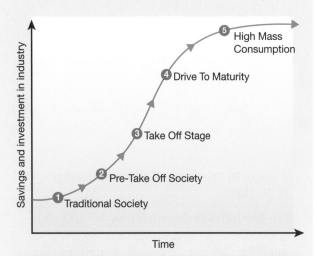

1 **Traditional society** – Most people work in agriculture in a 'subsistence economy'.
2 **Pre-conditions for take-off** – a shift from farming to manufacturing. Trade increases profits, which are invested into industries and infrastructure.
3 **Take-off** – rapid growth. Investment and technology create new manufacturing industries.
4 **Drive to maturity** – a period of growth. Technology is used throughout the economy. Industries produce consumer goods.
5 **Age of high mass consumption.** Societies choose how to spend wealth, e.g on military strength and education.

Figure 1 *Rostow's five-stage model of economic development*

Frank's dependency theory

In 1967, André Frank developed his dependency theory. He believed:

- development was about two types of global region – core and periphery (Figure 2)
- core regions were the developed nations
- periphery regions were the 'others', producing raw materials to sell to the core. They depended on the core for their market.

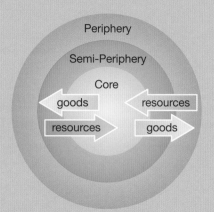

- Low-value raw materials are traded between the periphery and the core.
- The core processes these into higher-value products, and becomes wealthy.

Frank believed that:

- historical trade was what had made countries poor initially
- poorer countries are weaker members of a global economy.

Figure 2 *Frank's Core-Periphery model*

 Six Second Summary

- Rostow believed that countries should pass through five stages of development.
- Frank believed that countries are poor because of their relationships with other countries.

 Over to you

Copy Figures 1 and 2. Annotate **four** things on each to check you understand Rostow and Frank's models.

Student Book
See pages 64–65

You need to know:

- why some countries benefit from globalisation more than others.

What is globalisation?

GDP in sub-Saharan Africa lags behind the rest of the world, and Malawi struggles. Could globalisation help?

Globalisation means the ways that countries become increasingly connected. This happens through:

- economic **inter-dependence**
- increasing **trade**
- the spread of **technology**
- international **flows** of investment
- **outsourcing** services (e.g. call centres)
- the spread of media and **culture**.

Changes in investment

In the 1990s, US and European TNCs invested in south-east Asia.

- Investment by one country into another is called **Foreign Direct Investment** (FDI).
- TNCs could manufacture goods cheaply in China because wages were 90% lower than in the USA and Europe.
- The result was a huge growth in Chinese exports.
- It caused a **global shift** in manufacturing from developed to developing countries.

The Clark-Fisher model

This model explains three changes in employment structure as countries develop (Figure 1).

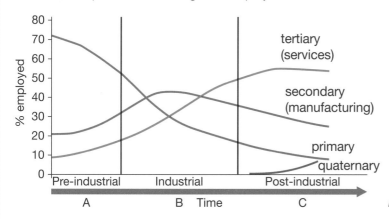

A Low-income countries – Employment is dominated by the primary sector.

B Middle-income countries are dominated by the secondary sector (**manufacturing**). As incomes rise, demand for manufactured goods increases.

C High-income countries – As incomes continue to rise, the tertiary sector (services) develops. The quaternary sector develops as tertiary services become more specialised, e.g. IT.

Figure 1 *The Clark-Fisher model*

Development in Vietnam

Since 1990, Vietnam has seen rapid **industrialisation**.

- Money has been invested in factories, turning raw materials (**primary products**) into manufactured goods (**secondary products**).
- While Vietnam has benefitted from globalisation, Malawi is stuck, and its GDP remains low (Figure 2).

	Malawi	Vietnam	UK
GDP per person (in US$)	900	3400	37 700
Where GDP comes from (%)	Agriculture: 30.3 Industry: 16.7 Services: 53	Agriculture: 22 Industry: 40.3 Services: 37.7	Agriculture: 0.6 Industry: 20.6 Services: 78.8

Figure 2 *Comparing the economies of Malawi, Vietnam and the UK in 2014*

 Six Second Summary

- Manufacturing has shifted from developed to developing countries.
- The Clark-Fisher model describes changes in employment structure as countries develop.
- Not all countries have benefited from globalisation.

 Over to you

Write ten questions about what you have learned so far in this chapter. Test yourself in a few days.

Student Book
See pages 66–67

You need to know:

- about India's growth, and its significance as a country.

It's all about growth!

India is one of the world's **emerging countries**, and is predicted to become the world's second largest economy by 2050.

- Its economy has nearly quadrupled in size since 1997.
- India's location has encouraged its economic growth (Figure 1).

- India is located between the Middle East and south-east Asia, the world's fastest growing economic regions.
- Half of the world's population lives in China, southern and south-east Asia, providing labour and a huge market for goods and services.

The Chinese economy has doubled in size every ten years! It's the world's largest exporter of manufactured goods.

The UAE economy doubles in size every 14 years! It has the world's largest GDP per capita.

The south-east Asian 'tigers' (Singapore, Hong Kong, Malaysia, Indonesia, Thailand and the Philippines).

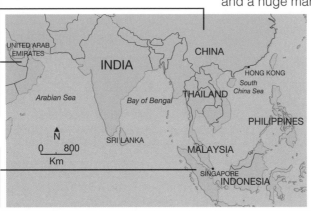

Figure 1 *The significance of India's location, politically and economically*

Understanding India's significance

Socially, India has:

- the world's second largest population, 1.3 billion in 2017!
- the world's 4th and 5th largest cities – Mumbai and Kolkata.
- some of the world's worst urban slums, housing 40 million people.

Environmentally, India has some of the world's:

- richest biodiversity, such as elephants and tigers. But population and economic growth threaten them.
- worst environmental problems, with land, air and water pollution. India is the world's third greatest emitter of greenhouse gases.

Politically, India is:

- the world's largest democracy.
- a growing global influence. It contributes the second-largest number of troops to the UN.

Culturally, India:

- is diverse. In 2011, 78% of the population practiced Hinduism, 15% Islam, 2.5% Christianity, and 2% Sikhism.
- has amongst the world's most ancient cultures.
- has the world's largest film industry, Bollywood.

 Six Second Summary

- India's economy is growing, fast, helped by its geographical location.
- It has the world's second largest population, and largest democracy.
- It is culturally diverse.
- It has rich biodiversity but environmental problems.

Over to you

Draw a table to show the social, political, cultural and environmental significance of India. Highlight in different colours positive and negative factors.

*Student Book
See pages 68–69*

You need to know:

- the reasons for rapid globalisation in India, and its impacts.

Globalisation and economic growth

The theory of globalisation is easy – if a country can't produce everything it needs, then it must trade.

- Companies always try to produce goods wherever it is cheapest.
- Countries such as India gain, with cheap labour.

The effects of globalisation on India (Figure 1) have been:

- exports increased by almost 20 times in 23 years
- increased output, as measured by total GDP.

The impacts on India's people have been:

- a 500% increase in GDP per capita
- reduced unemployment and poverty.

	India 1991	India 2014
GDP total, (US$) in PPP	1.2 trillion	7.3 trillion
GDP per capita (US$) in PPP	1150	5800
Exports value (US$)	17.2 billion	342 billion
Imports value (US$)	24.7 billion	508 billion
Unemployment rate %	20	8.6
Living in poverty %	36	30
Main exports	Commodities – tea, coffee, iron ore, fish products	Petroleum products, gems and jewellery, machinery, steel, chemicals, vehicles, clothing
Main imports	Petroleum products, textiles, clothing, machinery	Crude (unrefined) oil, gems and jewellery, machinery, fertilizer, iron ore

Figure 1 *Economic change in India 1991 – 2014*

Economic liberalisation in India

India's globalised economy began in 1991, with a programme of **economic liberalisation**. This means that consumers and companies (rather than the government) decide:

- what people will buy, based on demand
- where goods can be made most cheaply
- where investment in products will make most profits.

Governments supporting a market economy encourage foreign investment.

Foreign Direct Investment

Much of India's economic growth has come from Foreign Direct Investment (FDI).

- Most of it has come from major **Trans National Companies** (TNCs) and international banks.
- The service economy has grown most, with TNCs investing in IT, research and development, and call centres, all providing cheaper services.

The importance of transport

Three changes have reduced transport costs.

- **Shipping**. Improvements in fuel efficiency mean that a large ship costs only slightly more to run than a small one. Global shipping has increased hugely.
- **Containerization**. Containers on ships are easier, quicker and cheaper to transport.
- **Aircraft technology**. Only 0.2% of UK imports arrive by air, but these make up 15% by value. Imports from India by air are 70 times more valuable (e.g. jewellery) than those transported by sea.

Six Second Summary

- Globalisation has increased India's exports and output.
- Economic liberalisation has encouraged Foreign Direct Investment.
- Shipping, containerization and aircraft technology have reduced transport costs.

Over to you

Create a mind map to show the reasons for globalisation in India, and its impacts.

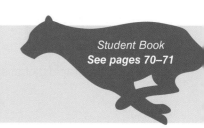

Student Book
See pages 70–71

You need to know:

- how one TNC (BT) operates in India, including outsourcing.

It's all change!

In India fewer people now work as farmers, while services have seen a huge growth in value (Figure 1).

	1991	2014
Agriculture % of GDP	31	18
Manufacturing % of GDP	28	24
Services % of GDP	41	58

Figure 1 *How contributions to India's GDP have changed*

The part played by TNCs in India's growth

- In India, companies such as BT have led **outsourcing** – where a company moves services overseas because labour is cheaper.
- India benefits because a high proportion of its qualified population speak English.

Three types of outsourcing have occurred in India:

- **Call centres**. Most Indian call centre employees are graduates earning 20% of what BT has to pay in the UK.
- **Software development**. Indian universities provide technically-qualified graduates.
- **Company administration**, e.g. accounting.

BT – a global company (Figure 2) – has Indian headquarters in New Delhi, though software development takes place in Bangalore – India's 'Silicon Valley'.

Figure 2 *BT, a global company, has offices around the world*

- The Indian government offers reduced taxes to attract companies.
- Changes in communications technology help companies like BT. Conference calls make it easy to 'meet' virtually instead of in person.

BT and the 'new' economy

BT is typical of TNCs that are part of the 'new economy' – based on the sale of services, rather than manufactured products.

- These companies are '**footloose**', as long as high-quality communication links are available.

- The new economy is also known as the '**knowledge economy**' because it relies on skilled, well-qualified people.
- TNCs usually locate close to universities and science parks.
- Most decisions are made at global headquarters, with other operations in lower-wage countries.

 **Six Second Summary**

- Outsourcing is where a company moves services overseas, where labour is cheaper.
- Three types of outsourcing have occurred in India.
- BT is part of the 'new economy' or 'knowledge economy'.

 Over to you

Summarise how TNCs and outsourcing have helped India to develop.

Student Book
See pages 72–73

You need to know:

- the social and economic impacts of change in India.

Head for the city!

- In 1990, 25% of India's population was **urban**.
- By 2015, this had risen to 33% – 400 million people!
- The increase in jobs has led to **rural-urban migration** – the movement of people to live in cities.

A time of social change

Economic development has social impacts, including urbanisation.

- **Urban expansion**, particularly the construction of new apartments.
- For educated women, a career results in **later marriage**. Birth and fertility rates fall (Figure 1).
- **Population structure** changes – with a lower dependency ratio.
- **Social customs** change. In cities, young urban Hindus are freer to marry outside their caste.

Changes in health and education have occurred (Figure 2). India's falling infant mortality rate is due to:

- increased access to safe water reducing waterborne diseases
- rapid expansion of hospitals in rural areas.

Population Indicators	1991	2014
Birth rate (per 1000 people)	30	19.9
Death rate (per 1000 people)	10	7.35
Fertility rate	4.0	2.5
% population aged 0–14	37.7	28.5
% population aged 65 and over	3.8	5.8
Dependency ratio	70.4	51.8

Figure 1 *Changes in India's population data, 1991 – 2014*

HDI, Health and Education Indicators	1991	2014
HDI	0.38	0.59
Life expectancy (years)	59.7	68
Infant mortality per 1000 live births	89	43.2
Maternal mortality per 100 000 births	550	200 (2010)
Number of doctors per 100 000 population	41	70 (2012)
Average number of years in school	2.4	12 (2011)
Literacy rate %	50	74 (2011)
Average age of first marriage for women	18.7	20.2 (2009)

Figure 2 *Changes in HDI, health (green) and education (blue) data for India, 1991 – 2014. Women's literacy remains 17% lower than men's.*

Economic change – Winners and losers

India's middle class is growing, but that still leaves a billion who are not well paid.

Garment workers

By 2015, clothing was India's largest manufacturing industry.

- It earned US$ 300 billion in GDP.
- India's minimum wage is 87% lower than in the UK.
- There is no shortage of people willing to work 100 hours a week in factories, for an average of £35.
- Most textile jobs are unskilled. 70% of employees are young women on lowest pay, and older women are often discriminated against.

Looking forward – can growth last?

- By 2050, India's population will be ageing.
- If birth rates fall further, its dependency ratio will rise.

Six Second Summary

- An increase in urban jobs has led to rural-urban migration.
- Economic development has social and economic impacts.

Over to you

Highlight the social and economic impacts of change in India in different colours. Close your book and list them.

Student Book
See pages 74–75

You need to know:

- how change has affected different parts of India in different ways.

Booming India – but for whom?

India's wealth is concentrated in cities.

- People migrate to cities and spend money on housing and services, creating more jobs.
- This causes the **multiplier effect** (Figure 1).
- Over time, this effect can create a **core region**.
- Wealth is unequally shared and other regions don't reap the same benefits.

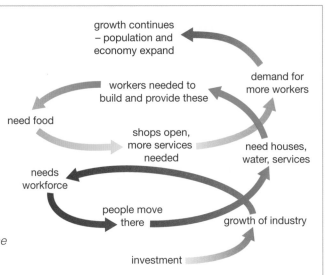

Figure 1 *The upward spiral caused by the multiplier effect*

Maharashtra – an urban core region

Maharashtra is the state with the biggest GDP (Figure 2) and it contains India's largest city, Mumbai.

Maharashtra's economic growth has come from:

- **service industries** – e.g. banking, IT, call centres
- **manufacturing** – e.g. clothing and engineering
- its **port** – the second largest in the country
- a booming **construction** industry
- **entertainment** – Mumbai hosts the Bollywood film industry.

	Per capita income, rupees
Maharashtra	104 000
Bihar	27 000
India, average	68 000

Figure 2 *Per capital income in rupees. 100 Rupees = £1*

The environmental impacts of growth

India's economic growth has made environmental issues worse including:

- water pollution from waste removal, and sewage
- air pollution from traffic, and coal-fired power stations
- loss of biodiversity and land degradation, as more land is needed for food and cities.

Bihar – the rural periphery

The state of Bihar is a part of India's **periphery**.

- It receives little investment, and is distant from cities.
- 86% of its population is rural. Many are subsistence farmers.
- People can't afford basic services – only 59% have electricity.
- School attendance is low and many are illiterate.
- Most women are low-wage labourers.

 Six Second Summary

- Maharashtra's economic growth has created an urban core region.
- Bihar is part of the rural periphery; many people are subsistence farmers.
- Economic growth has environmental impacts.

Over to you

Explain the differences in total GDP between Maharashtra and Bihar.

A top-down project:
The Narmada River Scheme

You need to know:

- about top-down development in India.

Student Book
See pages 76–77

Top-down – the government decides!

Rainfall over India is unevenly spread.

- Between November and March, almost no rain falls.
- Between May and September, the **monsoon** brings heavy rain.
- Dams make it possible to store water for the dry season.

The Indian government has built over 4500 dams, 14 of which are **super dams**. These can:

- encourage economic development by providing drinking water and electricity
- make it possible to farm dry lands, using **irrigation**.

Top-down development and IGO funding

Top-down development involves:

- decision-makers – usually governments or large companies who identify needs
- experts who plan changes.

Local people have no say in whether, or how, the plans happen.

Top-down schemes:

- are usually large and expensive,
- often involve loans from **Inter-Governmental Organisations** (IGOs) – i.e. government banks and agencies.

For example, the Sardar Sarovar Dam has been funded by the World Bank, Japanese banks, and Indian state governments.

The argument is that everyone benefits from 'top-down' because jobs and wealth 'trickle down' to the poor.

The Sardar Sarovar Dam

The Sardar Sarovar Dam, on the Narmada River (Figure 1), is one of the world's largest dams.

Who benefits from the dam?

- **India's cities** – from the provision of drinking water and hydroelectric power (HEP).
- **Farmers** in western India – from irrigation water.

Who loses?

- **Local residents and farmers** – 234 villages have been flooded by the dam; rural communities can't afford the electricity; farmland has been flooded; fertile sediment is no longer deposited on flood plains.
- **Western India** – religious and historic sites have been flooded.

- **People downstream** – the weight of large dams could trigger earthquakes, destroying communities and causing flooding.

Figure 1 *The Sardar Sarovar dam*

 **Six Second Summary**

- The Sardar Sarovar Dam is an example of top-down development.
- It is funded by state governments, banks and the World Bank (an IGO).
- The dam has positive and negative impacts.

 Over to you

Create a revision card for the Sardar Sarovar Dam, which outlines the: **a)** characteristics; **b)** advantages; **c)** disadvantages, of the top-down strategy.

A bottom-up project: biogas

Student Book
See pages 78–79

You need to know:

- about bottom-up development in rural India.

Working from the bottom up

Not all development projects are like the Sardar Sarovar Dam (Section 2.13).

- In many cases, experts work with communities to identify needs and offer help.
- This is known as **bottom-up development**.
- It's usually run by **non-governmental organisations** (NGOs), e.g. charities or universities.

The ASTRA project

ASTRA (Application of Science and Technology in Rural Areas) is a development project in rural India. By talking to families it found three problems.

1 Most rural girls have little **education**.
2 Daily routine – including collecting firewood – left little **time** for school.
3 There is less wood to use for **fuel** for cooking.

Solution – think cow dung!

The answer to all three problems was cow dung!

- Cow dung produces **biogas**. This is used for cooking by day, and powering electricity generators at night.
- Dung is fed into a pit that forms part of a biogas plant (Figure 1).
- It ferments to produce methane which is piped into homes.
- It's simple, uses local materials, and is an example of **intermediate technology**.

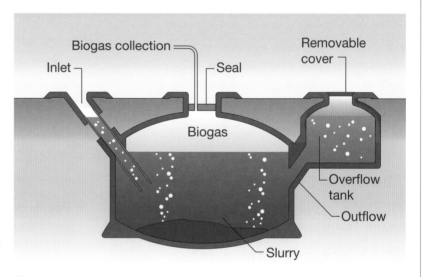

Figure 1 *A biogas plant*

The benefits of biogas

By 2010, four million cattle dung biogas plants had been built in India.

- These created 200 000 permanent jobs, most in rural areas.
- Cooking with gas is smoke-free, so there are fewer lung infections.
- Girls have more time to go to school as they are not gathering wood.
- Sludge produced in the biogas plant is rich in nutrients so it makes a good fertiliser.
- Many villages use biogas to power electricity generators to pump water from underground, so farmers can get three crops a year.

 Six Second Summary

- Biogas plants are an example of bottom-up development, which involves working with communities.
- Biogas plants have brought economic and social benefits, especially for girls.

 **Over to you**

List the benefits of biogas and classify them as economic, social or environmental. Some fit into more than one category!

India – which way next?

Student Book
See pages 80–81

You need to know:

- about challenges that India faces and how its role in the world is changing.

The challenges ahead

Despite rapid economic growth, India has not invested enough in its **infrastructure**.

- 25% of people have no electricity supply.
- Per capita electricity consumption is low and the electricity network is inadequate (Figure 1).

There are problems with water, education and wealth distribution.

- Water resources are declining. Projects such as the Narmada schemes (Section 2.13) are essential, if controversial.
- Too few people have a good education.
- Poverty is widespread (Figure 2). Little wealth reaches rural villages.

300 million Indians earn less than US$1 a day. 45% of Indian children under-five are malnourished. Two-thirds of India's homes have no toilet.

Figure 2 India's future problems – adapted from an article on the BBC World Service online

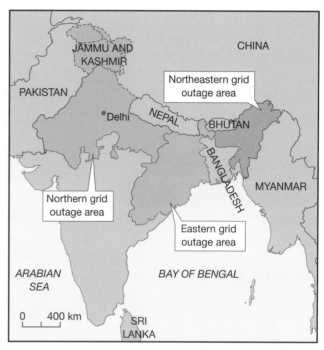

Figure 1 Areas of India affected by a major power blackout in July 2012 which affected 50% of India's (and 9% of the world's) population

What's the cause of India's problems?

- To attract investment, India's tax rates on businesses are low. So TNCs pay little tax.
- For those on high incomes, there are many ways to avoid paying tax.
- The government therefore has relatively little income to pay for public services, e.g. water, sanitation.

Six Second Summary

- India has not invested enough in its own infrastructure.
- Its government does not receive enough tax revenue to pay for public services.
- As India develops, its role is increasing in Asia and globally.

Over to you

Write down your top ten key facts from this page.

India on the world stage

As one of the world's largest economies, India's role is increasing in Asia, and also globally.

- In **Asia** relations with some neighbours are tense. There have been three wars with Pakistan since 1947. Kashmir, lying between them, remains disputed (see Figure 1).
- Water scarcity can cause conflict. The main rivers of Pakistan and India rise in Kashmir's mountains.
- Indian HEP projects could take irrigation water from farming areas of Pakistan.

- **Globally**, India belongs to the G20 group of the world's largest economies, which includes the USA and some EU members.
- Through the G20, India can help resolve global problems e.g. climate change.
- India can also now support investment from the World Bank and development banks, to help the economies of developing countries.

Topic 2 – Development dynamics   49

Topic 3
Challenges of an urbanising world

Your exam

- Topic 3 Challenges of an urbanising world makes up Section C in Paper 1, Global geographical issues.

- Paper 1 is a 90-minute written exam and makes up 37.5% of your final grade. The whole paper carries 94 marks (including 4 marks for SPaG) – questions on Topic 3 will carry 30 marks.

- You have to answer all questions in Paper 1. Section A contains questions on Hazardous Earth (pages 12–33), and Section B on Development dynamics (pages 34–49).

Tick these boxes to build a record of your revision

Your revision checklist

Spec Key Idea	Detailed content that you should know	1	2	3
3.1 The world is becoming increasingly urbanised	• Past, current, and future global trends in urbanisation, how it varies between regions			
	• The global pattern of megacities and their influence			
3.2 Urbanisation is a result of socio-economic processes and change	• How economic change and migration contributes to the growth / decline of cities in different countries			
	• Why urban economies differ in developing, emerging and developed countries			
3.3 Cities change over time and this is reflected in changing land use	• How urban population numbers, distribution and spatial growth change over time			
	• Characteristics of different urban land uses and the factors that influence land-use type			
3.4 The location and context of the chosen megacity influences its growth, function and structure	• Site, situation and connectivity of the megacity in a national, regional and global context			
	• The megacity's structure in terms of its functions and building age			
3.5 The chosen megacity is growing rapidly	• Reasons for past and present trends in population growth			
	• How population growth has affected the pattern of spatial growth, changing urban functions and land use			
3.6 Rapid population growth creates opportunities and challenges for people living in the megacity	• Opportunities for people living in the megacity			
	• Challenges for people living in the megacity caused by rapid population growth			
	• Residential pattern of wealth and slums and squatter settlements, and the challenges of managing the megacity			
3.7 Quality of life in the megacity can be improved by different strategies for achieving sustainability	• Advantages and disadvantages of top-down strategies for making the megacity more sustainable			
	• Advantages and disadvantages of bottom-up strategies for making the megacity more sustainable			

Student Book
See pages 84–85

You need to know:

- about past, current, and likely future trends in urbanisation.

Urbanisation

Urbanisation occurs as people move from rural to urban areas.

How does urbanisation vary between different regions?

In 2007, for the first time, more people lived in urban areas than rural.

- The biggest increase is in Asia and Africa (Figure 1).
- **Asia's** urban population is expected to grow to 64% by 2050.
- **Africa's** urban population will grow to 58% by 2050. This will still be the world's lowest urban percentage.

The causes of this growth are:

- migration to cities
- natural increase, i.e. more births than deaths.

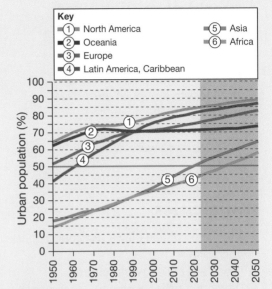

Figure 1 *The increasing urban population in different world regions, 1950–2050 (projected)*

The changing balance

Big changes have taken place in the distribution of the world's ten largest cities (Figure 2):

- In 1975, six were in developed countries. These cities grew during the industrial revolutions of the 18th and 19th centuries.
- Urbanisation in developing countries has mostly taken place since the 1950s, and urban populations double every 30 years!
- Even so, less than 40% of people in developing countries lived in urban areas in 2015.
- By 2025, only two of the ten largest cities will be in developed countries. Urban populations here are now rising slowly.

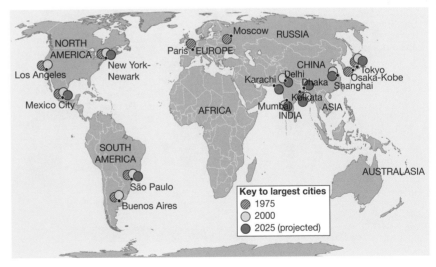

Figure 2 *The ten largest cities in the world*

Six Second Summary

- Urbanisation is the rise in the percentage of people living in urban areas.
- Asia and Africa are expected to see the biggest rises in urban population.
- Most of the world's largest cities are now in developing countries.

Over to you

Use a voice recorder to record your understanding of past, current, and likely future trends in urbanisation. Replay your recording to help you to remember it.

*Student Book
See pages 86–87*

You need to know:

- about the global pattern of megacities and world cities.

A world of millionaires

When urbanisation occurs, towns and cities grow in population and area.

- The term **million city** is used for any city with a population of over one million.
- In 1950, there were 83 million cities.
- By 2015, there were over 500!

The growth of megacities

Megacities have over 10 million people.

- In 1980, most megacities were in high income countries – New York, Tokyo, Paris, London. The populations of some of these have hardly changed since 1950, but a few (like London) are now rising fast.
- By 2015, 75% of the world's megacities were in emerging countries, e.g. Sao Paulo, Shanghai and Mumbai.

World cities

- A few megacities play a disproportionate role in world affairs.
- These are called **world cities**.
- They have **urban primacy** – that means an importance and influence bigger than their size suggests. London is one of these.

Think of each world city as a wheel. The cities are 'hubs' (centres), where economic activity occurs. Spokes radiate out with flows of investment, airline traffic, decision-making and political decisions.

In 2012, the world cities were graded based on their influence in the global economy (Figure 1).

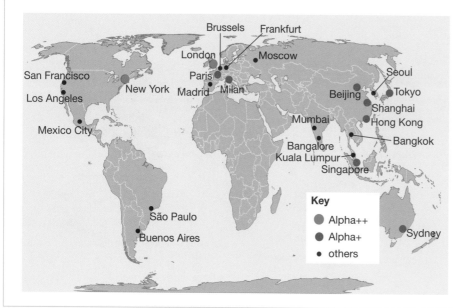

- **Decision-makers**. 80% of the world's largest companies have headquarters in cities of the USA, the EU, and Japan.

- **Investment**. London and New York are the world's biggest financial centres.

- **Airline traffic**. In 2014, Dubai was the world's largest international airport. But add together London's airports (e.g. Heathrow, Gatwick), then London is the biggest.

- **Political decisions**. Government decisions in the UK can affect people globally, e.g. about where to invest, or trying to resolve conflicts.

Figure 1 The major world cities

 Six Second Summary

- Megacities have over 10 million people.
- Increasing numbers of megacities are in emerging countries.
- Cities with urban primacy have an influence bigger than their size suggests.

Over to you

Create a table to show how some urban areas have disproportionate economic and political influence.

*Student Book
See pages 88–89*

Urbanisation on a huge scale!

The world's fastest-growing cities are in Asia and Africa. The main cause of growth is economic growth, which creates new jobs.

- In emerging countries, TNCs and manufacturing have caused rapid industrialisation.
- In high income countries (HICs), some 'world cities' are growing rapidly as their service economies expand.

In each case, migration causes urbanisation as people move to find work.

Case Study 1 – Kampala

Kampala is the capital of Uganda.

- Its growth is driven mainly by **internal migration**, but also by natural increase.
- **Rural-urban migration** is a result of factors which 'pull' people to Kampala, and others that 'push' them from the countryside (Figure 1).

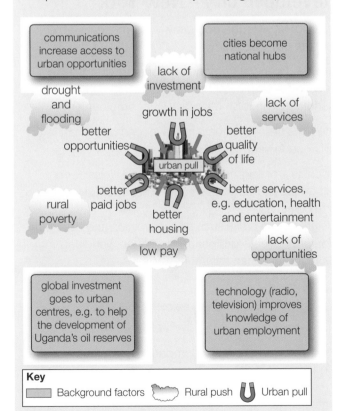

Key

⬜ Background factors 　☁ Rural push 　🧲 Urban pull

Figure 1 Rural-urban migration in Uganda

Case Study 2 – New York

The population growth of many cities in HICs has slowed. But New York's has been speeding up. Its growth has come from:

- **net growth** from overseas migration
- natural increase
- the **knowledge economy**. New York's focus on finance has increased **international migration**.

Case Study 3 – Detroit

Detroit, USA, is home to General Motors, the world's largest vehicle company. The city has experienced population decline.

- In 1950 the population was 1.85 million, but only 673 000 in 2017.
- Between 1960 and 2000, its wealthier population left, leaving a poorer population.
- Income from local taxes reduced and services declined.
- Between 2000–10 General Motors sales halved. It now makes cars using robotics, needing fewer people.
- It buys parts from overseas, putting local supply companies out of work, leading to **de-industrialisation**. People have left to find work elsewhere.

⏱ Six Second Summary

- Kampala is growing largely because of rural-urban migration.
- New York's knowledge economy attracts international migrants.
- De-industrialisation has led to population decline in Detroit.

✏ Over to you

For each case study, create a flow diagram to show how economic change has led to changes in the urban populations.

Student Book
See pages 90–91

You need to know:

- how urban economies differ in developing, emerging and developed countries.

The informal economy

Traders in the **informal economy** do not figure in most development data – yet millions earn their living on the street, selling goods or offering a service.

A developing city – Kampala

Kampala's informal economy is large, and Uganda earns half of its estimated GDP from informal work (Figure 1).

- Most informal workers are women and young people, and are poor.

The **formal economy** is growing slowly because most Ugandans are subsistence farmers.

- Manufacturing employs only 5% of Uganda's population.
- Services are the main part of Kampala's formal economy e.g. shops and offices.

Figure 1 *A fruit and vegetable stall on the street in Uganda*

An emerging city – New Delhi

New Delhi is wealthy.
Even so, street selling is common.

- India's 'Hindu' newspaper suggests that 75% of New Delhi's workers are in the informal economy.
- Services earn New Delhi 78% of its GDP.
- Manufacturing contributes 20%, but the clothing industry is growing fast.
- Much of the informal economy is in factories but there are no regulations about minimum wage, benefits or working conditions.

A developed city – New York

New York's knowledge economy is the most valuable part of the city's economy.

- In 2014, financial companies alone provided 10% of New York's employment.
- Manufacturing is small with 10% of employment.

Economists claim that the informal economy earns 7% of US GDP each year.

- It consists of migrants, both legal and illegal, and self-employed workers who may not declare income to tax officials.
- It is greatest in areas such as construction and catering.
- Workers have no protection, and often work long hours for less than minimum wage.

 Six Second Summary

- Kampala's informal economy is large.
- Much of New Delhi (and India's) informal economy is in factories, where there are few regulations.
- The most valuable part of New York's economy is its knowledge economy, but it also has a sizeable informal economy.

 Over to you

Compare three things about how urban economies differ in developing, emerging and developed countries.

Student Book
See pages 92–93

Why New York grew

New York began in the 17th century as a fort on the island of Manhattan (Figure 1).

• The city's deep harbour enabled it to trade with Europe, and was the main entry point for immigration.
• Irish migrants came to escape famine in the 1840s.
• Millions from eastern and southern Europe arrived in the 1870s and 80s.
• Communities formed **ethnic enclaves**.

Figure 1 Four boroughs of New York city: Manhatten, Brooklyn, The Bronx, and Queens

Suburbanisation

Manhattan soon became crowded. The **subway** and **rail** system expanded, making urban expansion possible.

• From Manhattan, people could go to The Bronx, Brooklyn or Queens (Figure 1).
• After the 1930s, **road bridges** fed traffic into Manhattan on **freeways** from Long Island and The Bronx.

Counter-urbanisation and 'white flight'

• From 1950–1980, New York lost 12% of its population due to **counter-urbanisation** (Figure 2).
• Those who left tended to be white second-generation migrants.
• This 'white flight' left behind poorer migrant communities and Black Americans.
• Income from taxation fell. By 1975 New York was nearly bankrupt.

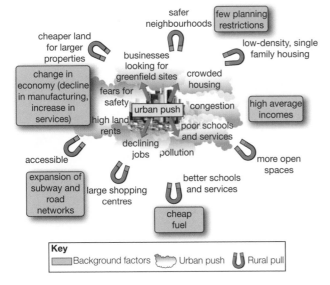

Figure 2 Reasons for New York's counter-urbanisation

Re-urbanisation

Since 1980, three changes have attracted people back into New York, known as **re-urbanisation**.

• The knowledge economy created jobs.
• Closure of docks and industries has created space for **regeneration** with new apartments and offices on **brownfield sites**.
• The city is safer due to 'zero tolerance' policies towards crime.

Six Second Summary

• Transport made suburbanisation possible in New York.
• Counter-urbanisation involved mostly white, wealthier people – it was known as 'white flight'.
• The knowledge economy and regeneration have encouraged re-urbanisation.

Over to you

Define the following terms:

• urbanisation
• suburbanisation
• counter-urbanisation
• re-urbanisation.

Student Book
See pages 94–95

You need to know:

- about different urban land uses and what causes these.

Understanding cities

Land use in cities is usually arranged in a pattern (Figure 1), and is easy to recognise.

- City centres (commercial areas) look different from residential areas (where people live).
- Each of these looks very different from industrial areas.

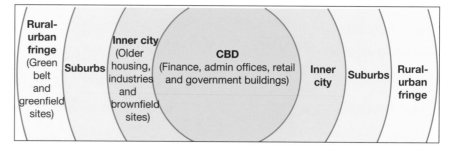

Figure 1 *The pattern of land use in towns and cities*

The table below shows how the pattern of land use has developed. It's caused by:

- **accessibility** (how easy a place is to get to)
- **cost of land**
- **planning decisions** made mostly by councils.

Type of land use	Characteristics	How these areas develop
Commercial (offices and retail)	• Mostly in the CBD – **Central Business District**. • Taller buildings at a higher density than the rest of the city. • Also 'low-rise' business and retail parks on the city edge (the **rural-urban fringe**).	• The CBD is the most accessible part of the city. • Demand for land is greatest. • Space is limited – the only way to build is up! • Land is expensive – every bit of land is used. • Business parks are near main roads for easy access.
Industrial	• Either in the **inner city** (older 19th century industries) or on the city edge (more recent industries). • Close to transport links.	• Away from the CBD because they need space. • Older industries relied on canals, rivers and rail. New industries rely on road transport.
Residential	• Usually in **suburbs**. • The oldest properties are close to the centre. Housing varies between different parts of cities: • 19th century houses are **terraced**. • 20th century houses are often **semi-detached** and **detached**. • 21st century housing varies from apartments in the inner city to large housing estates on the outskirts.	• Cities grow in 'rings', with oldest suburbs near the centre and newest on the outskirts. • Land is expensive near the centre, so terraces and flats are common. Further away from the city, cheaper land means houses can have larger gardens. • In the 21st century, planners prefer to allow housing on 'brownfield' land rather than use 'greenfield' sites (farmland that has never been built on).

Figure 2 *Land use development*

 Six Second Summary

- Land use in cities is usually arranged in a pattern.
- Commercial, industrial and residential areas have different characteristics.
- Areas develop differently because of accessibility, cost of land and planning decisions.

 Over to you

- Draw a large copy of Figure 1.
- Use Figure 2 to add simple drawings of the types of buildings found in each zone.
- Annotate your diagram with reasons for the differences in each zone.

Mumbai – a growing megacity!

Student Book
See pages 96–97

You need to know:

- about the location of Mumbai and its urban structure.

Mumbai – world city

Mumbai is a **megacity**, India's main commercial city, and a world city. About 25 million people lived in the metropolitan area in 2015.

Mumbai's site and situation

- Mumbai lies by the estuary of the Ulhas River (Figure 1). Its port has grown round the estuary.
- During monsoon season, torrential rains flood low-lying roads.
- It has spread to form a conurbation, including Navi Mumbai, Thane, Bhiwandi and Kalyan.

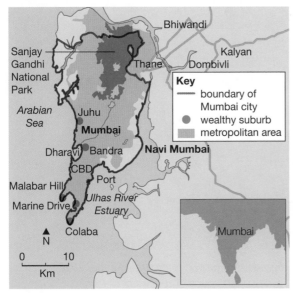

Figure 1 Mumbai's location around a natural harbour

National and international connections

- Mumbai's deep-water harbour has made it India's second biggest **port,** and its largest container port.
- Its west coast location makes it closer to Europe (via the Suez Canal) than other Indian ports.
- By **air**, Mumbai is nine hours or less from UK, Singapore and Middle East destinations.
- Most Indian cities are within two hours' flight time.

The structure of Mumbai

Mumbai's structure is not *exactly* like the model shown in Figure 2.

- Because of the harbour, the CBD is not in the centre, but near the island tip (Figure 1).
- Industrial areas are near the port, or places such as Navi Mumbai (where land is cheaper).

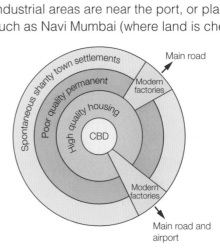

Figure 2 A model of the structure of developing cities

Residential areas in Mumbai show wide inequality.

- Wealthy suburbs (Figure 1) are all inner city areas along waterfronts, close to the CBD.
- Middle-low income areas are in older parts of the city, further from the CBD.
- Low-income groups live in **'chawls'** – low quality multi-storey buildings.
- The poorest 60% live in **informal** housing.
- Thousands live on Mumbai's streets.

Six Second Summary

- Mumbai is on an estuary, where the port grew.
- Mumbai is well-connected, both nationally and internationally.
- Its structure isn't typical of developing cities. Its CBD is near the island tip, and there is wide inequality in residential areas.

Over to you

Write down each title from this page. Summarise your understanding of each section, using three bullet points per title.

Geographical skills

You need to know:

- how to identify changes in Mumbai and its spatial growth, using maps and photographs.

*Student Book
See pages 98–99*

The development of Bombay / Mumbai

Before 1995, Mumbai was known by the name Bombay, which dated from the time when India was a British colony (this ended in 1947).

- In 1995, the Hindu nationalist party won elections and announced that Bombay would be renamed after the Hindu goddess Mumbadevi.
- In this section, you can trace **spatial** changes in Mumbai as it grew between 1888 and 2015.

 Big Idea

The spatial growth of a city means how much extra space it takes up as it grows.

Figure 1 Bombay in 1888

Figure 2 High density growth in central Mumbai 2015 – there is now very little open space. The area of the whole city is now 10–15 times the area shown in this aerial photo!

 Six Second Summary

- Historic maps and images can be used to investigate spatial growth.
- Spatial growth means how much extra space a city takes up as it grows.
- Mumbai has grown considerably between 1888 and 2015.

 Over to you

- **a)** Describe Bombay as a settlement in 1888 (Figure 1) – where the original settlement was, its location, and the economic activities there.
- **b)** Using the scale, measure the size of the settlement N-S and E-W, and then calculate its area in km².
- **c)** Estimate what percentage of the total map area was built up in 1888.
- Estimate what percentage of the area of central Mumbai shown in Figure 2 is built up.
- Describe the environment of central Mumbai shown in Figure 2. Use Google Earth or Google Maps if you need more detail.
- Using Figure 2, suggest reasons why some middle class people are leaving central Mumbai to live in metropolitan suburbs.

Student Book
See pages 100–101

A thousand a day

About 1000 new migrants arrive in Mumbai every day.

- Mumbai's population is growing by 3% a year – this rapid growth is called **hyper-urbanisation**.
- The population of Mumbai was about 16 million in 2015 and will reach 20 million by 2020.
- By 2050 it will probably be the world's largest city.

Mumbai's growth and economy
Pattern of spatial growth

- Mumbai has expanded in area and population.
- New suburbs, such as Navi Mumbai, are growing, caused by the migration of the middle classes from the city.
- 60% of Mumbai's population lives in slum suburbs (Figure 1).

Figure 1 Dharavi, in Mumbai, is a slum suburb

Changing investment and land use

Investment has grown, increasing the amount of employment. Investment has been greatest in:

- services
- manufacturing
- construction
- entertainment and leisure.

Mumbai's growth puts pressure on land and it is one of the world's most expensive cities. Many car manufacturers – e.g. Audi and Skoda – are moving out of Mumbai because they need large amounts of land.

Over to you

Create two mind maps to show: **a)** the causes; **b)** the consequences of population growth in Mumbai.

The growth of Mumbai

Mumbai has grown for two main reasons.

1 Rural-urban migration

- Maharashtra receives most migrants because it is India's wealthiest state.
- Rural migrants head for the biggest cities – Mumbai, Delhi, Kolkata (Figure 2).

Mumbai offers:

- jobs
- education facilities
- entertainment
- higher incomes

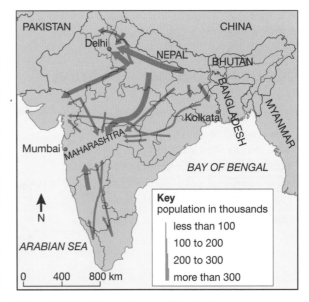

Figure 2 Migration patterns within India

2 Natural increase

- Migrants tend to be in their 20s and 30s.
- Those who find work usually settle and start families.
- Mumbai's natural increase is 1.4% per year – nearly half the city's annual growth.

Six Second Summary

- Mumbai is experiencing hyper-urbanisation.
- Mumbai's population has grown due to rural-urban migration and natural increase.
- Population growth has created new suburbs, and increased slum suburbs.

Student Book
See pages 102–103

You need to know:

- about inequalities in lifestyles within Mumbai.

Economic opportunities

Many in Mumbai work in the **informal economy** doing jobs like rag-picking (Figure 1).

- There's no regular wage, contract, job security, health and safety protection or pension scheme.
- But the informal economy adds US$1 billion to Mumbai's GDP each year.

Many who work in Mumbai's informal economy live in slums.

- Dharavi is the largest slum and quality of life is poor (Figures 2 and 3).
- But Dharavi is cheap, and for many it's convenient for work.

Figure 1 *Rag-picking – making money from other people's rubbish – is an important part of the informal economy in Mumbai*

Life at the bottom – Dharavi

The shanty houses were so close that daylight would hardly reach the pathways. Up to 10 families might share a tap – if water was flowing (which it often didn't) and toilets were communal, costing 3 cents to use.

Figure 2 *Adapted from 'A walk in Dharavi', by Jim Yardley, New York Times, 2011*

Figure 3 *2015 Dharavi factfile*

People	
Population of Dharavi	Estimated 800 000–1 million
Area	2.39 km² (the size of London's Hyde Park)
People per home	Between 13 and 17
Average size of home	10 m² (equivalent to a medium-sized bedroom)
Hygiene and health	
People per individual toilet	625
Most common causes of death	Malnutrition, diarrhoea, dehydration, typhoid
Education	
Literacy rate in Dharavi	69% (Mumbai average is 91%)

Life in the middle

[Adapted from news articles]

The Chopra family (2 adults, 2 children) live in a small flat. They all sleep in one room. Mr Chopra's teaching salary is not enough, so he tutors privately, increasing his monthly salary to 75 000 rupees (£750).

India's middle class is growing and their incomes rising. Mr Chopra says, 'Nearly every family has a TV and mobile phone. My children are growing up in a different world.'

Life near the top

Vihaan, 25, has a degree from Mumbai's top engineering college, speaks English, and is part of Mumbai's upper middle class. He earns 1.6 million rupees (£16 000) a year.

His company provides an apartment. In 2015, high-spec one-bedroomed apartments in Mumbai cost £320 000. Vihaan is saving to buy one himself.

Six Second Summary

- There are many inequalities in lifestyles within Mumbai.
- Many who work in Mumbai's informal economy live in slums.
- India's middle class is growing and their incomes rising.

Over to you

Create a table to show how employment, housing, income and lifestyle vary in Mumbai.

*Student Book
See pages 104–105*

You need to know:

- about the challenges facing Mumbai caused by its population growth.

Problems ahead?

Mumbai has become India's economic giant. Its industries – like the port – are its **formal** economy.

- However, **employment conditions** vary for many people.
- Most of Dharavi's factories are illegal; many are sweatshops.
- These are **informal** – low pay, no security, and no tax.
- Tax is the problem – without tax income Mumbai's government can't provide services for the population (Figure 1).

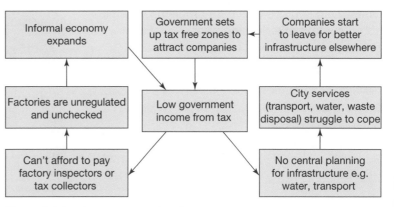

Figure 1 A low-tax government cycle

Housing shortages and slum development

Mumbai's population is growing so rapidly, there's a housing shortage.

- City authorities have no money to build housing.
- Private companies are put off building because the government limits rents.
- Many live in cramped rooms, far from work.
- Others squat on streets or spare land.
- Over time, they find materials to build a home. Once there, few can afford to move.

Water supply and waste disposal

60% of Mumbai's population uses communal taps. In some slums, water only runs for 30 minutes a day.

- Factories use the Mithi River to dump untreated waste.
- 800 million litres of untreated sewage go into the river every day.
- However, 80% of Mumbai's waste is recycled.
- The recycling industry is worth US$1.5 million a year, and employs 10 000 people.

Air pollution and traffic

In 2015, an air quality index was introduced to improve air quality in Mumbai. Suggestions for improvement include:

- using LPG instead of burning coal
- improving public transport
- charging higher road tax on older vehicles.

Little is spent on transport infrastructure. Roads and trains are overcrowded, and 3500 people die on Mumbai's railway each year.

Six Second Summary

Challenges facing Mumbai include:

- too little tax to improve conditions
- housing shortages and slum developments
- uncontrolled water pollution
- air pollution and traffic congestion.

Over to you

Draw a spider diagram to explain how a lack of tax income leads to other challenges for Mumbai. Develop your explanations.

You need to know:

*Student Book
See pages 106–107*

- whether top-down strategies can make Mumbai more sustainable.

Sustainability and the future

One way to improve Mumbai's problems is to think about **sustainable development**. Sustainability can be measured in two ways (Figure 1):

1 as a **'stool'**. If economic, social and environmental benefits outnumber problems, then it's a good idea.
2 as a **'quadrant'**. Ideally, the answer to all four questions, is 'yes'.

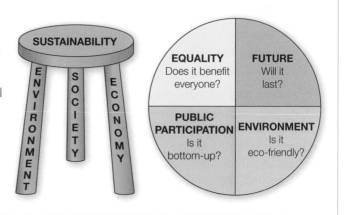

Figure 1 *Two ways of measuring sustainability – the 'three legged stool' and the 'quadrant'. Both are just as useful.*

Top down development – 'Vision Mumbai'

'Vision Mumbai' is a plan for **top-down development** to improve the city and quality of life. It has ambitious targets (Figure 2).

Figure 2 *Vision Mumbai's four core targets*

Putting plans into action

Some things could be done quickly, e.g.:

- restoring 325 'green' spaces that were polluted and used for dumping waste
- building 300 extra public toilets
- widening main roads
- improving train capacity and safety.

But the main plan was based on property development.

- Dharavi would be demolished.
- Instead, high-rise blocks for slum dwellers would be built next to shopping malls, offices and luxury apartments.

Has Vision Mumbai worked?

Improvements brought by Vision Mumbai	Objections by Dharavi residents to Vision Mumbai
<ul><li>By 2007, 45 000 homes were demolished in Dharavi.</li><li>New flats replaced slums.</li><li>Piped water and sewerage systems were established for new flats.</li><li>By 2015, 72 new trains were introduced on Mumbai's railways.</li><li>New measures were introduced to improve air quality (see Section 3.11).</li></ul>	<ul><li>Many prefer slum improvement (e.g. piped water, sewage treatment) to demolition.</li><li>Apartment blocks have split communities.</li><li>Rents cost more than in the slums - many can't afford them.</li><li>Small workshops would have to move, affecting Mumbai's recycling industry.</li></ul>

Figure 3 *Has Vision Mumbai worked?*

In spite of the plans:

- water quality in Mumbai is worsening because of sewage discharge.
- improving sewage treatment and disposal in line with population and slum growth is a long way off.

Six Second Summary

- 'Vision Mumbai' is a top-down development.
- It has advantages and disadvantages for people living in Dharavi.

Over to you

Highlight in different colours the references to managing water supply, waste disposal, transport and air quality (both advantages and disadvantages). Test yourself in three days.

Student Book
See pages 108–109

You need to know:

- about bottom-up ways of making Mumbai more sustainable.

What happens when you're ill in Mumbai?

In India, you either pay for health care, or use insurance. But better health would improve sustainability.

- Some charities focus on health issues.
- These are Non-Government Organisations (NGOs), and normally work in communities on **bottom-up** development schemes.

 Big Idea

Bottom-up development is where experts work with communities to identify their needs, offer help, and let people have more control over their own lives.

LSS – a case study of a health charity

Lok Seva Sangam (LSS) is a health charity working in Mumbai. It was set up to control **leprosy** (Figure 1) on the edge of Dharavi.

LSS does a number of jobs such as:

- surveying communities to detect skin diseases
- setting up dermatology (skin) clinics
- running pharmacies to dispense drugs for treatment
- running kindergartens for young children, to help parents.

Figure 1 *The effects of leprosy. Many people are affected like this over their entire body.*

1 Health-related work

- In 30 years, LSS treated 28 000 people for leprosy. 75% were cured.
- It has now expanded its work to treat people with tuberculosis (TB).
- In 2015, LSS employed full-time medical staff, and volunteers.

2 Education

LSS now works mainly in another slum – Baiganwadi – focusing on education. The teachers' work:

- surveys and detects cases of TB and leprosy

- educates people about symptoms
- treats people by persuading them first that treatment is easy.

3 Community work

LSS also works with communities to teach about health. Much of its work is with women, focusing on:

- **sanitation** – e.g. boiling drinking water, and waste disposal.
- education about **vermiculture** (worms composting waste). Worms reduce the bacteria in household waste. The compost produced can then be sold.
- **activities** e.g. paper bag making, to aid discussion as well as making items to raise money.

 Six Second Summary

- LSS is a health charity providing bottom-up development.
- It delivers education about health.
- It also carries out health-related work and community work.

 Over to you

List ten key facts to remember about bottom-up work carried out by LSS.

Topic 4
The UK's evolving physical landscape

Your exam

- Topic 4 The UK's evolving physical landscape makes up Section A in Paper 2, UK geographical issues.

- Paper 2 is a 90-minute written exam and makes up 37.5% of your final grade. The whole paper carries 94 marks (including 4 marks for SPaG) – questions on Topic 4 will carry 27 marks.

- You must answer all parts of Section A. Section B contains questions on The UK's evolving human landscape (pages 88–108), and Section C contains questions on fieldwork, where you have a choice of questions (pages 109–129).

Tick these boxes to build a record of your revision

Your revision checklist

Spec Key Idea	Detailed content that you should know	1	2	3
4.1 Geology and past processes have influenced the physical landscape of the UK	• Role of geology, tectonic and glacial processes in upland and lowland landscapes			
	• Characteristics and distribution of sedimentary, igneous, and metamorphic rocks			
4.2 A number of physical and human processes work together to create distinct UK landscapes	• Processes forming distinctive upland and lowland landscapes			
	• Why distinctive landscapes result from human activity			
4.3 Distinctive coastal landscapes are influenced by geology interacting with physical processes	• How geological structure and rock type influence erosional landforms			
	• How climate and erosional processes create coastal landscapes and retreat			
	• How sediment transport and deposition influence coastal landforms			
4.4 Distinctive coastal landscapes are modified by human activity interacting with physical processes	• How human activities affect coastal landscapes			
	• How physical and human processes change one named coastal landscape			
4.5 Human and physical processes present challenges along coastlines, with a variety of management options	• Why there are increasing risks from coastal flooding			
	• Costs and benefits of managing coasts using hard and soft engineering, and more sustainable approaches			
4.6 Distinctive river landscapes have different characteristics formed by interacting physical processes	• Channel size and shape along upper, middle and lower courses of rivers			
	• Erosion, transport and depositional processes in river landform formation			
	• How climate, geology and slope processes affect river landscapes			
	• Storm hydrographs			
4.7 River landscapes are influenced by human activity interacting with physical processes	• How human activities affect storm hydrographs			
	• Processes leading to flooding on one UK river			
4.8 Some rivers are more prone to flood, with a variety of river management options	• Risks and threats of river flooding			
	• Costs and benefits of managing floods using hard and soft engineering			

Landscapes from the past

You need to know:

- how geology and past processes have created upland landscapes in the UK.

Student Book
See pages 112–113

Explaining the past

Malham Cove in the Yorkshire Pennines is a spectacular UK upland area which was once a huge waterfall higher than Niagara Falls. It is formed of limestone, made up of crushed coral shells. So how did this get to be 300m above sea level?

The landscape around Malham Cove results from three factors, summarised in Figure 1.

Geology	• **Carbon dating** has proved that fossils in the limestone at Malham Cove lived in the tropical seas that covered the UK during the Carboniferous period. • The limestone was formed by compaction of skeletons of marine organisms. • Other strata were deposited on top of the limestone, e.g. sandstone and shale. • The highest peaks consist of rocks most resistant to **erosion** (Figure 2).
Tectonic processes	Over 300 million years ago, tectonic processes affected the Pennines. 1. The plate on which the UK sits moved away from the tropics. 2. Convection currents uplifted rocks from beneath the sea to become land and caused rocks to snap, tilt and move along **faults**. Some parts were raised more than others, forming a fault scarp as at Giggleswick Scar.
Glaciation	The glaciers of the most recent Ice Age (10000 years ago) caused: 1. River valleys to deepen and widen (Figure 2). 2. Features such as Malham Cove when they melted.

Figure 1 *Factors in the formation of the landscape around Malham Cove*

Big Idea

The UK's upland and lowland landscapes are the result of geology, tectonic processes and glaciation.

Upland and lowland landscapes

- Upland areas of the UK consist of resistant igneous, metamorphic and some sedimentary rocks.
- Lowland areas of the UK generally consist of younger, less resistant sedimentary rocks.

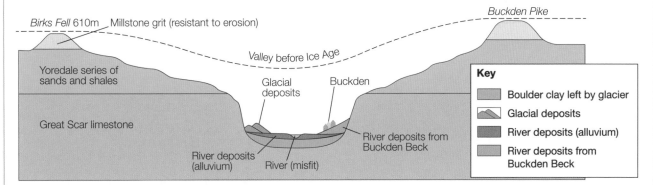

Figure 2 *Geological cross section of Wharfedale, a valley in the Pennines. Notice the U-shaped valley deepened and widened by a glacier.*

Six Second Summary

- The landscape around Malham Cove results from its geology, past tectonic and glacial processes.
- The limestone at Malham Cove was created in the Carboniferous period when the UK was covered by tropical seas.
- The Pennines were affected by three tectonic processes.
- Past glaciers changed the shape of river valleys and left features like waterfalls.

Over to you

Name **a)** three ways in which rock types influence the Pennine landscape, **b)** three tectonic processes which have affected the Pennines, and **c)** two effects of the most recent Ice Age.

You need to know:

- about the relationship between landscape and geology
- the characteristics and distribution of rock types in the UK.

Student Book
See pages 114–117

The shape we're in

The height and shape of the UK's landscape depends on the rocks from which it is formed.

Geology has played a role in the economy:

- Cornwall's tin and copper made it wealthy
- coal helped make the UK the world's first industrial nation.

 **Big Idea**

There is a direct link between geology and the shape of the UK's landscape.

How do rocks differ?

There are three main types of rock (Figure 3).

- **Igneous** – the oldest rocks, formed from lavas and deep magmas. Most are resistant to erosion.
- **Sedimentary** – formed from sediments eroded and deposited by rivers or the sea. Some are resistant, others crumble easily.
- **Metamorphic** – sedimentary rocks that were heated and compressed during igneous activity. They are hard and resistant.

Rocks and landscape

Relief (landscape) depends greatly on rock type. The 'Tees-Exe line' on Figure 1 joins the River Tees in north-east England with the River Exe in the south-west. Notice that:

- west of the line is mainly upland
- east of the line is mainly lowland.

Now look at Figure 2. Notice that north and west of the Tees-Exe line:

- most rocks are older
- the most resistant igneous and metamorphic rocks are found
- there are more faults; these areas were uplifted by tectonic activity.

East of the line, most rocks are:

- younger
- weaker sedimentary rocks, which erode easily. The limestone here is younger and less resistant than Carboniferous limestone.

Figure 1 *UK relief map*

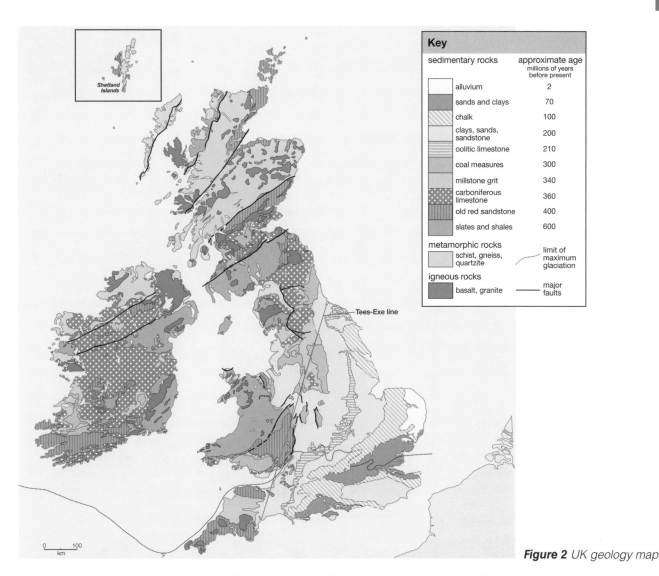

Figure 2 *UK geology map*

Igneous	How it was formed	Characteristics
Granite	From magma cooling deep underground.	Contains crystals of quartz, feldspar and mica. Very resistant.
Basalt	From lavas rich in metals.	Almost black, and heavy. Very resistant.
Sedimentary		
Chalk	A purer, younger form of limestone.	Very porous. Medium resistance.
Carboniferous limestone	From compacted skeletons of marine animals	Permeable. Generally resistant.
Clay (shale when compacted)	From muds deposited by rivers or at sea.	Soft and crumbly. Generally weak.
Sandstone	From sand grains compacted together.	Slightly porous. Weak (less than 100 million years old); and resistant (over 300 million years old).
Millstone grit	From cemented and compacted sandstone.	Very resistant.
Metamorphic		
Slate	From heated muds or shale.	Very resistant.
Schist	From further metamorphosis of slate, where it partly melted and solidified.	Very resistant.
Marble	From heated limestone.	Very resistant.

Figure 3 *Ten rocks you should know!*

 **Six Second Summary**

- There are three main rock types: igneous, sedimentary and metamorphic.
- The different rock types formed in different ways, have different characteristics and are located in different parts of the UK.
- The UK's landscape (relief) depends largely on rock type.

 **Over to you**

Make flashcards for each of the three rock types. Give an example of each one.

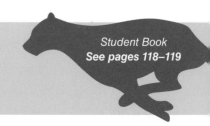

You need to know:

- how upland and lowland landscapes result from different physical factors.

Student Book
See pages 118–119

An upland landscape

The Lake District in north-west England is the UK's wettest region, and colder than southern England. The landscape has been affected by physical processes.

Weathering and slope processes

Rainwater gets into cracks in rocks. It freezes and expands when temperatures fall below 0° C, and thaws when temperatures rise. This freeze-thaw **weathering** breaks up rock creating **scree** (angular rock pieces) which collect at the foot of slopes.

Slope processes affect valley sides.

- Scree is unstable and moves easily during **rockfalls**.
- Rain adds weight to the weathered rock, causing **landslides**.

Post-glacial river processes

The Lake District was once **glaciated**. Glaciers created deep U-shaped valleys. Today, small rivers (**misfits**) deposit silt and mud (**alluvium**) in these large valleys. This makes them fertile for farming.

A lowland landscape

The Weald, in Kent and Sussex, was once a continuous arch called an **anticline**. Erosion has left bands of resistant rock alternating with less resistant rock creating **undulating** (gently rolling) hills (Figure 1). This is **scarp and vale topography**.

Weathering

Southern England is warmer than the Lake District, so different types of weathering occur (see Section 4.16), including:

- **chemical weathering** – e.g. solution of chalk
- **biological weathering** – e.g. tree and shrub roots break up solid rock.

Post-glacial river and slope processes

During, and after, the last Ice Age, water in the porous chalk froze, making it impermeable allowing

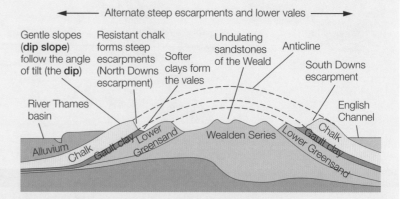

Figure 1 *A cross section of the geology of the Weald*

rivers and valleys to form. As the climate warmed, the ice thawed and water seeped through the chalk leaving **dry valleys**. Clay is impermeable so rivers are common in vales.

Slope processes are slower than in the Lake District. The most common is **soil creep**, caused by rain dislodging soil particles (see Section 4.16).

 Six Second Summary

- Weathering, slope processes and post-glacial river processes create different features in upland and lowland landscapes.
- Weathering is the physical, chemical or biological breakdown of solid rock.
- The Lake District is an upland landscape with high peaks, U-shaped valleys and misfit rivers.
- The Weald is an undulating lowland landscape with scarp and vale topography.

 Over to you

Explain the differences between physical (freeze-thaw), chemical and biological weathering.

Student Book
See pages 120–121

You need to know:

- how distinctive landscapes result from human activity such as agriculture, forestry and settlement.

Living among the trees

- Every landscape has been shaped by the original settlers – their activities and the way they made use of materials around them.
- Settlements can develop for specific reasons. For example, Ae in Scotland was created for workers who planted new woodlands, to replace wood used in the trenches in World War I.

Big Idea

Human activities have shaped the UK's landscape.

The Yorkshire Dales

Norse farmers and settlers of the 8th and 9th centuries left their mark on the Yorkshire Dales.

- Limestone and rocks left by melting glaciers (see Section 4.1) made excellent building stone.
- Land was cleared for farming, and the stone used as field boundaries.

Cold winters and a short growing season influenced farming.

- Sheep grazed upland fells and were brought to the valley in winter (as happens today).
- Winter feed was stored in stone barns, often in longhouses (a house and barn). This led to a dispersed pattern of isolated farms.

Figure 1 *Settlers in the Yorkshire Dales used local materials to create dry stone walls and buildings*

East Anglia

East Anglia is low-lying and generally flat. European Angles (hence its name) and Vikings settled there in communal villages.

- The surface geology is mainly glacial sands and clays (**till**) from the last Ice Age. Till produces fertile soil for arable (crop) farming, but little solid stone for building.
- Below the surface, the solid geology is chalk. Chalk is too crumbly for building, but contains pieces (**nodules**) of flint, a hard crystalline form of quartz. Many older buildings were built from this.

Figure 2 *With little building stone, field boundaries are marked by hedges or ditches*

Six Second Summary

- An area's landscape is influenced by people who settled there – through their economic activity and use of local materials.
- In the Yorkshire Dales, limestone is used for walls and farm buildings.
- In East Anglia, hedges or ditches mark field boundaries because there is little solid building material.

Over to you

Draw a table to show the influence of **a)** geology, **b)** human activity and **c)** settlement on the landscapes of the Yorkshire Dales and East Anglia.

Student Book
See pages 122–123

You need to know:

- how geological structure and rock type influence coastal erosional landforms.

Geology and rock type

Geology is the main influence on the characteristics of the **coastal zone**.

- **Hard rock coasts** consist of resistant rock, such as igneous granite, or sedimentary chalks (e.g. in Lulworth Cove in Dorset).
- **Soft rock coasts** consist of less resistant rock, such as sedimentary sands and shales (e.g. in Christchurch Bay in Dorset/Hampshire).

 **Big Idea**

Coastal landscapes are influenced by the interaction of geology and coastal processes.

Rock structure

Rock structure means how rock strata (layers) are arranged, for example:

- at right angles to the coast (**discordant**)
- parallel to the coast (**concordant**).

Discordant coasts: headlands and bays

South-west Ireland is a discordant coast where resistant sandstones alternate with less resistant limestones (Figure 1). Waves have eroded limestone to form bays, leaving headlands of harder sandstone.

Concordant coasts: coves and cliffs

The coast at Lulworth Cove in Dorset is a concordant coast with unique geology (Figure 2).

Weaknesses in rock structure

Weakness in rocks influence coastal erosion. These are:

- **joints** – these are small, usually vertical, cracks
- **faults** – these are larger cracks caused by past tectonic movements.

Rocks with joints and faults are more easily eroded – a joint widens to form a **cave**; erosion creates an **arch** which collapses to form a **stack**. Eventually, this becomes a **stump**.

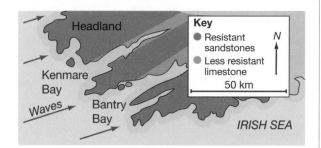

Figure 1 *The discordant coastline of south-west Ireland*

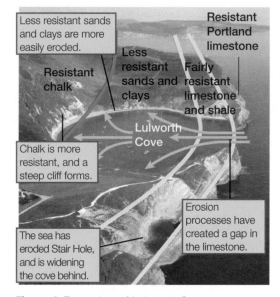

Figure 2 *Formation of Lulworth Cove on a concordant coastline*

 **Six Second Summary**

- Geology is the main factor influencing the coastal zone.
- A discordant coast is characterised by headlands and bays.
- Erosion of concordant coasts can produce coves.
- Joints and faults are weaknesses which can be eroded.

Over to you

Explain the differences between concordant and discordant coasts.

The UK – climate and coastline

4.7

You need to know:

- how climate and erosion processes influence the shape of coasts.

Student Book
See pages 124–125

What causes waves?

Friction between wind and water surfaces causes waves. Wave size depends on:

- wind strength and duration
- the **fetch**.

Prevailing winds in the UK (those which blow most often) are from the south-west, so south-west facing coastlines feel the full force of autumn and winter storms.

Beach profile (shape)

How waves break determine the beach profile.

In summer, waves arrive slowly, with long wavelengths (the distance between them) and low amplitudes (or height); these are **constructive** or **spilling** waves.

In winter, strong winds result in waves with large amplitude and short wavelength; these are **destructive** or **plunging waves**). They arrive quickly.

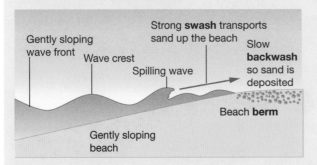

Figure 1 *Summer constructive waves build up a beach*

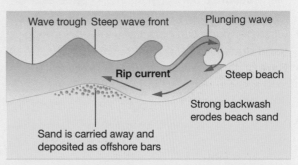

Figure 2 *Winter destructive waves erode beaches*

Coastal erosion

On coasts with resistant rocks, erosion is slow. Most happens during winter storms (Figure 3).

- **Abrasion** forms a wave-cut notch at the cliff base.
- As the notch grows, a cliff overhang develops.
- The overhang collapses. The resulting debris protects the base from further erosion.
- The rock debris is eroded, exposing the cliff to erosion again.

This causes the cliff line to retreat. A level area of smooth rock is left where the cliff line once was (called a **wave-cut platform**).

1. Hydraulic action
Water is forced into cracks. When the wave retreats the compressed air blasts out. This can force the rock apart.

2. Abrasion
Loose rocks are thrown against the cliff by waves, wearing away and chipping off bits of rock.

Cliff

Waves crashing against cliff

3. Attrition
Sediment constantly collides and gets worn down into smaller, rounded pieces.

Figure 3 *The three main types of coastal erosion*

Six Second Summary

- Waves are caused by wind blowing over water surfaces.
- Wave size is determined by wind strength, duration, and fetch.
- In summer, constructive waves create gently sloping beach profiles.
- In winter, destructive waves create steep beach profiles.
- Abrasion, hydraulic action and attrition are types of coastal erosion.

Over to you

Memorise definitions for the words in bold on this page.

Test yourself tomorrow.

Student Book
See pages 126–127

You need to know:

- how coastal processes create depositional landforms.

Sediment transport

- Material eroded from cliffs is called **sediment** (see Section 4.7). It can be transported to new locations.
- The main way in which sediment is transported is by **longshore drift** (Figure 1).

Depositional landforms

If eroded sediment is trapped in sheltered areas (e.g. bays), a beach forms. Other sediment transported by longshore drift can create new landforms where it is deposited (Figure 2).

- Many **beaches** consist of sand and pebbles moved gradually by longshore drift.
- Onshore winds can blow sand inland, forming **sand dunes**.
- **Bars** of sand can extend across a bay (due to longshore drift). Behind the bar, a shallow **lagoon** forms.
- Where longshore drift meets an estuary, the river flow stops the drift and sand is deposited forming a **spit.**
- At high tide, the sea flows inland causing the spit to curve back on itself (a **recurved** end).
- The water behind a spit is sheltered, allowing **salt marshes** to form.

Stabilising sand dunes

As sand dunes develop, plants stabilise them, because they:

- have long roots to hold them in place in strong winds
- have tough, waxy leaves to resist sandblasting
- can survive being sprayed by salt water.

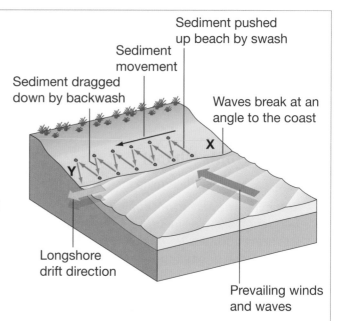

Figure 1 Longshore drift can transport sediment hundreds of kilometres before it is deposited

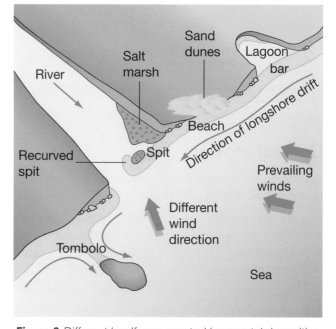

Figure 2 Different landforms created by coastal deposition

Six Second Summary

- Sediment results from cliff erosion.
- Longshore drift transports sediment along the coast.
- Beaches, spits, bars and sand dunes form where sediment is deposited.
- Vegetation can help stabilise depositional landforms.

Over to you

Draw a flow diagram to show how longshore drift forms a spit.

You need to know:

- how development, agriculture, industry and coastal management affect coastal landscapes.

Student Book
See pages 128–129

The effects of development

Development can put pressure on a crowded coastline. It includes:

- Tourism – with demands for hotels and campsites in tourist areas.
- Housing – many London workers cannot afford housing there, so commute from coastal areas. Many retire to coastal areas.
- Offices and industry. The high cost of London property makes coastal locations popular with companies (e.g. bankers JP Morgan moved to Bournemouth in 2015).

The effects of agriculture

Romney Marsh in Kent is a wetland habitat for birds, and summer grazing pasture. It faces two pressures.

- The high cost of good farmland means farmers have to make best use of whatever extra grazing they can. This reduces natural wildlife habitats.
- Climate change and rising sea levels may lead to salt water floods during winter high tides, threatening pastures.

The effects of industry

Bacton gas terminal (Figure 1), on the Norfolk coast is next to a sandy beach. The beach is a popular tourist destination but being next to a gas terminal is not ideal. Similar industrial developments have taken place next to the rivers Solent (Southampton), Thames (London), Severn (Bristol), Mersey/Dee (Liverpool) and Tees (Middlesbrough) estuaries.

Figure 1 *Bacton gas terminal on the Norfolk coast*

 **Big Idea**

Human activities affect coastal landscapes.

The effects of coastal management

Milford-on-Sea in Christchurch Bay in Hampshire has suffered from coastal erosion. Coastal management (Figure 2) is intended to prevent further problems. (Also see Section 4.12.)

Wooden groyne | Stone groyne | Recurved sea wall

Figure 2 *Milford-on-Sea. Coastal management has affected the coastline by trapping sediment.*

 **Six Second Summary**

Human activities put pressure on coastal landscapes:

- Development (housing, offices and tourism) causes expansion of coastal towns and cities.
- Agriculture – wildlife habitats are threatened by demand for grazing; pastures are also threatened by flooding due to climate change.
- Industry – large industrial developments cause conflict with tourists.
- Coastal management alters the landscape.

 **Over to you**

Look at Figure 2. How would the coastal management measures shown affect this and other coasts in the area?

You need to know:

*Student Book
See pages 130–131*

- how rising sea levels increase the risk of flooding and erosion on coastlines.

Rising sea levels

Climate scientists believe that global warming is causing ice sheets to melt and sea levels to rise. This puts low-lying areas at risk, such as Bangladesh, and the south-eastern UK.

Big Idea

A sea level rise of 1m by 2100 would have major impacts on people in low-lying areas.

Flood risks and the future

Sea levels constantly change (Figure 1):

- twice a day with high tides, due to the gravity of the moon
- twice a month with exceptionally high tides (called spring tides)
- due to falling air pressure, which causes storm surges. They are more severe when they coincide with spring tides and large waves.

Global warming may result in more frequent and deeper low pressure weather systems, causing higher storm surges.

- In December 2013, high winds and a 7 m storm surge struck eastern and south-east England causing the worst coastal flooding since 1953.

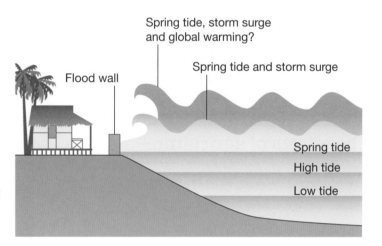

Figure 1 *People will become increasingly vulnerable to spring tides and storm surges as sea levels rise due to global warming*

What of the future?

The future holds several challenges:

- Events like the 2013 storm surge could become more frequent by 2050.
- Beaches, spits, and river deltas may be eroded faster, and become submerged.
- A sea level rise of 50 cm would make sea defences in south and east UK useless. The only choice would be to build higher defences or abandon some areas to the sea.

Six Second Summary

- Global warming may cause sea levels to rise by up to 1m by 2100.
- Many low-lying areas are at risk from rising sea levels.
- A combination of spring tides, increased storm surges and rising sea levels put more people at risk from flooding.
- Higher sea levels, increased storm activity, storm surges and erosion of the coast could make sea defences useless.

Over to you

Create a spider diagram with 'Flood risk' at the centre. Add legs to summarise the factors that result in a risk of coastal flooding.

*Student Book
See pages 132–133*

You need to know:

- why coastal erosion rates vary
- why cliffs collapse, and the impacts this has on people.

The rate of coastal erosion

Different factors determine the rate at which coasts erode.

- **Geology**. Cliffs of resistant rocks withstand waves for long periods. For example, Dodman Point in south Cornwall has suffered little erosion in 3000 years. Meanwhile, weak geology causes rapid erosion, e.g. the Holderness Coast in East Yorkshire.

- **Cliff processes** (Figure 1). Areas of weak geology suffer from **cliff foot erosion** (caused by hydraulic action and abrasion) and **cliff face erosion** (caused by **sub-aerial processes** e.g. weathering and movement of materials downslope, called **mass movement**).
- **Waves**. Wave energy depends on the fetch over which the wave has travelled (see Section 4.7).

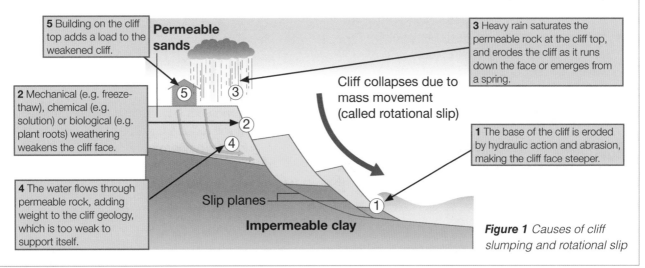

5 Building on the cliff top adds a load to the weakened cliff.

Permeable sands

3 Heavy rain saturates the permeable rock at the cliff top, and erodes the cliff as it runs down the face or emerges from a spring.

Cliff collapses due to mass movement (called rotational slip)

2 Mechanical (e.g. freeze-thaw), chemical (e.g. solution) or biological (e.g. plant roots) weathering weakens the cliff face.

1 The base of the cliff is eroded by hydraulic action and abrasion, making the cliff face steeper.

4 The water flows through permeable rock, adding weight to the cliff geology, which is too weak to support itself.

Slip planes

Impermeable clay

Figure 1 Causes of cliff slumping and rotational slip

Christchurch Bay

Managing erosion at Christchurch Bay on the UK's south coast is costly and difficult.

- Without it, cliffs would erode by over 2m a year, threatening residential areas such as Christchurch, Barton-on-Sea and Milford-on-Sea.
- At Barton-on-Sea, mass movement is the major problem, caused by weathering and water movement.
- Cliff foot erosion also plays a part, as the Atlantic fetch of 3000 miles brings big waves.

Impacts of erosion

Erosion in Christchurch Bay affects many people:

- Homeowners lose their homes to the sea. House values fall, and insurance is impossible to get.
- Rapid cliff collapses are dangerous for people on the cliff top, and on the beach.
- Roads and other infrastructure are destroyed.
- Erosion makes the area unattractive for tourism.

Local people argue that they need defences to protect the coast. But they are expensive, and there is no agreement about which works best (see Section 4.12).

 Six Second Summary

- The rate at which coastlines erode varies.
- Geology, cliff processes and waves influence the rate of erosion.
- Cliffs collapse due to marine processes, sub-aerial processes and human actions.
- Cliff erosion at Christchurch Bay poses threats to many people.

Over to you

Classify the processes and actions in the text boxes in Figure 1 into:

- marine foot processes
- sub-aerial cliff-face processes
- human actions.

You need to know:

- what coastal management is
- how hard engineering protects the coast.

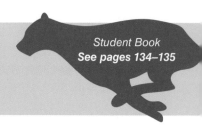

Student Book
See pages 134–135

 Big Idea

The method of coastal management used depends on environmental and economic **costs** versus the **benefits** of what is saved (**cost–benefit analysis**).

Managing the coast

There are two types of sea defences:

- 'hard' engineering – the traditional method uses concrete and steel structures to stop waves. It is expensive and unnatural/ugly.
- 'soft' engineering – uses smaller structures, often built from natural materials, to reduce wave energy.

Type of defence	Cost/ metre	Benefits and problems
Sea wall	£2000	• Reflect waves back out to sea. • Can prevent easy access to the beach. • Suffer from wave scour (plunging waves attack the wall's foundations).
Sea wall with steps and bullnose	£5000	• Steps help to dissipate wave energy; the bullnose throws waves up and back out to sea.
Revetments (sloping structures at top of beach)	£1000	• Break up incoming waves. • Restrict beach access; look ugly. • Can be destroyed by big storms.
Gabions (rock-filled cages)	£100	• A cheap type of sea wall. • Absorb wave energy (they are permeable). • Not very strong.
Rock armour (rip-rap)	£300	• Easy to build. • More expensive if built in the sea. • Dissipates wave energy; looks 'natural'.
Groynes	£2000	• Prevent longshore drift. • Larger beach **dissipates** wave energy, reducing erosion. • May increase erosion downdrift (Figure 2).

Figure 1 *Different hard defences and their costs*

Conflict in Christchurch Bay

The coast around Barton-on-Sea in Christchurch Bay is eroding rapidly (see Section 4.11).

Most people favour hard-engineering management, but some disagree.

- Coastal residents/businesses want to protect their interests.
- Locals living inland don't want to pay for expensive protection, preferring cheaper options.
- Environmentalists prefer a 'do nothing' or soft-engineering approach which has less impact on habitats/ecosystems.
- People downdrift worry that management updrift will reduce the size of their beaches (Figure 2). They prefer an integrated management plan for all affected.
- Politicians want protection but don't want to favour one group or another.
- Boat users want to protect access to the sea.

Highcliffe

Longshore drift

Beach grows

No new sediment reaches here

Rapid erosion

Stone groynes at Highcliffe trap sand

Beach cannot form

Figure 2 *The stone groynes at Highcliffe have built up the beach, but have caused rapid erosion downdrift. This is called* **terminal groyne syndrome**.

 Six Second Summary

- Coastal management includes hard and soft engineering.
- Hard engineering is costly and can be unattractive.
- A cost–benefit analysis helps to decide which type of sea defence should be used.
- Coastal management can lead to conflict.

 Over to you

Draw a table to show advantages and disadvantages of hard-engineering methods.

Student Book
See pages 136–137

You need to know:

- how coasts can be managed in a holistic, more sustainable way.

Big Idea

Sustainable (or 'soft') methods of managing coasts tend to work with the environment.

Managing the whole coast

Holistic coastal management means looking at the coastline as a whole, so actions in one area don't cause problems in another.

- It takes into account the needs of different people, the environment, costs and benefits. This approach is **Integrated Coastal Zone Management (ICZM)**.
- A **Shoreline Management Plan (SMP)** sets out how the coast as a whole will be managed.

Coastal management – the choices

There are four choices for coastal management:

1. **Hold the line** – sea defences stop erosion (expensive).
2. **Advance the line** – sea defences move the coast further into the sea (very expensive).
3. **Strategic realignment** (**strategic retreat**) – gradually let the coast erode.
4. **Do nothing** – take no action; let nature take its course.

Figure 1 shows choices about coastal management along the Norfolk coast.

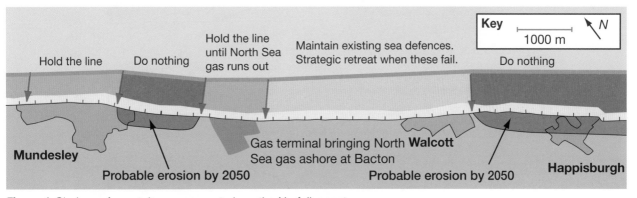

Figure 1 *Choices of coastal management along the Norfolk coast*

Soft engineering

Uses natural processes to stabilise beaches and cliffs, and reduce wave energy. It is often cheaper, less intrusive and more sustainable that hard engineering. Methods include:

- planting vegetation (£20 – £50/m²)
- beach nourishment: (£500 – £1000/m²)
- offshore breakwaters (£2000/m).
- cliff drainage (cost not available).

Planning for the future

The UK faces difficult decisions about how to protect the coast.

- It is too expensive to protect farmland and isolated houses.
- It is hard to persuade people that protecting their property is not sustainable.
- Planning defences is difficult if the impact of rising sea levels is uncertain.

Six Second Summary

- ICZM considers the needs of different people, the environment, costs/benefits.
- An SMP considers whole stretches of coast.
- Soft engineering uses natural processes to protect coasts.
- The UK faces difficult decisions about how best to protect the coast.

Over to you

Explain one advantage and one disadvantage of **three** methods of soft engineering.

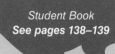

Student Book
See pages 138–139

You need to know:

- how to use your geographical skills to investigate coastal erosion in Christchurch Bay
- how to use 4- and 6-figure grid references and interpret an OS map.

Geographical skills

Christchurch Bay: key points

Barton-on-Sea has no protection from the prevailing south-westerly winds and waves. By studying the rate of erosion before, and after, coastal management (Figure 1), we can work out whether:

- erosion has changed as a result of coastal management
- money spent on managing the coast is justified.

This allows **stakeholders** (people affected by what happens to the coast) to judge whether coastal management is working.

 **Big Idea**

Map interpretation helps you to examine the impacts of coastal management and to predict possible effects of proposed schemes.

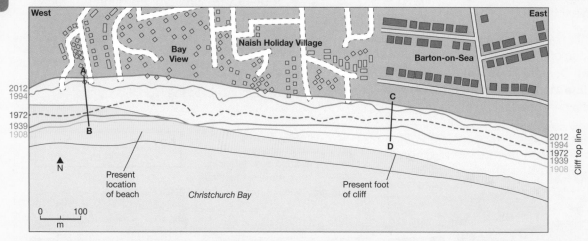

Figure 1 *Rates of erosion at two locations in Christchurch Bay, 1908–2012*

Figure 2 *1:25 000 Ordnance Survey (OS) map extract of part of Christchurch Bay*

 Six Second Summary

- In parts of Christchurch Bay erosion has been a problem for over a century.
- Studying the position of coastlines over time can help identify whether coastal management has affected the rate of erosion.
- It can also help stakeholders to decide whether money spent on managing the coast is justified.

Over to you

- Look at Figure 1. Use the scale bar to work out by how much the cliff top has eroded between 1908 and 2012 at both A–B and at C–D.
- On Figure 1, at location A–B, during which years was coastal erosion the greatest?
- Using Figure 2, compare the beach at High Cliff and the Hoburne Naish Holiday Park. How has coastal management affected them differently?

*Student Book
See pages 140–141*

You need to know:

- about river processes in upland areas.

Buckden Beck

The small rapids and waterfalls in Figure 1 are typical of a river's **upper course**. Because of the **gradient**, the stream flow looks fast, but it is actually flowing slowly because of friction with the stream bed.

Figure 1 *Upper course of Buckden Beck, a tributary of the River Wharfe in the Yorkshire Dales*

Erosion and transport

- The material carried by the river (the **load**) are the tools which erode its bed and banks (Figure 2).
- Most load is **transported** by solution, suspension, saltation and traction (Figure 3).

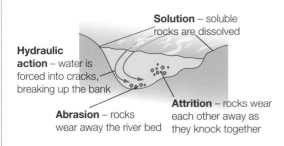

Figure 2 *River erosional processes*

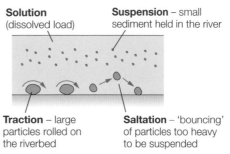

Figure 3 *River transport processes. Most load is carried during periods of wet weather.*

How does a waterfall form?

1. Waterfalls occur when a river crosses a bed of more resistant rock.
2. Erosion of the less resistant rock undercuts the hard rock above it. The river's energy creates a **plunge pool** at the foot of the waterfall.
3. Erosion of the less resistant rock creates a ledge, which overhangs and collapses.
4. The waterfall takes up a new position, leaving a steep valley or **gorge**.

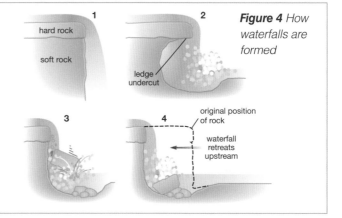

Figure 4 *How waterfalls are formed*

 Six Second Summary

A river in its upper course:

- has a steep gradient with small rapids and waterfalls
- flows slowly because of friction with the bed
- carries its load by solution, suspension, saltation and traction
- erodes its channel by abrasion, attrition, hydraulic action and solution.

 Over to you

Draw a table listing the processes of river transport in order of the size of the load carried. Include a brief description of each one.

Student Book
See pages 142–143

You need to know:

- how river valleys develop in upland areas
- how weathering and mass movement affect the shape of a river valley.

The valley of Buckden Beck

Figure 1 is typical of a river's upper course in an upland area.

- It has steep valley sides and a narrow bottom (**V-shaped**).
- As the river cuts vertically into the resistant carboniferous limestone, it winds around areas of more resistant rock. This produces **interlocking spurs** – ridges that jut into the valley from both sides.
- Physical, biological, and chemical **weathering** of the valley sides are important in shaping the valley (Figure 1).

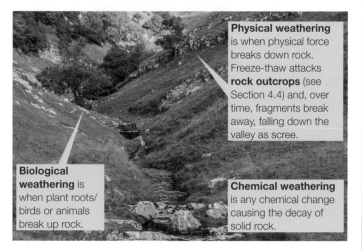

Physical weathering is when physical force breaks down rock. Freeze-thaw attacks **rock outcrops** (see Section 4.4) and, over time, fragments break away, falling down the valley as scree.

Biological weathering is when plant roots/ birds or animals break up rock.

Chemical weathering is any chemical change causing the decay of solid rock.

Figure 1 *The V-shaped valley and interlocking spurs of Buckden Beck, and weathering processes*

Mass movement

This is the movement of weathered fragments downslope by gravity. It can be rapid (e.g. **landslides** and **mudflows**) or slow (e.g. **soil creep**). The effects of soil creep are much less obvious than rapid movement. But over many years, it can cause walls, telegraph poles and trees to lean.

The shape of the valley

Valley shape is affected by three things (Figure 2):

- the rate of weathering
- the rate of mass movement
- how quickly the river can remove material. If the river has the energy, it takes the material and uses it to erode the valley, making it steeper. If flow is slow, weathered rock collects at the bottom of the slope, making the valley gentler and flatter.

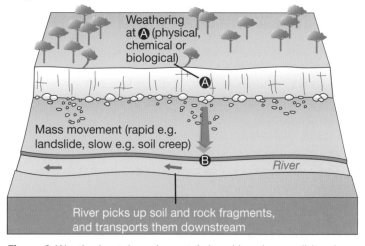

Weathering at **A** (physical, chemical or biological)

Mass movement (rapid e.g. landslide, slow e.g. soil creep)

River

River picks up soil and rock fragments, and transports them downstream

Figure 2 *Weathering takes place at A, breaking down solid rock into fragments. Mass movement moves the fragments down slope to the river at B. The river uses them as tools to wear away (erode) the river bed.*

Six Second Summary

- The valley in a river's upper course is V-shaped with interlocking spurs.
- Physical, biological and chemical weathering takes place on the valley sides.
- Mass movement moves weathered material downhill.
- Valley shape is affected by rates of weathering and mass movement, and how quickly a river can remove weathered material.

Over to you

Draw a sketch to show the features of a valley in a river's upper course.

Student Book
See pages 144–145

You need to know:

- how rivers and valleys change in the middle course.

The River Wharfe in its middle course

Figure 1 shows the River Wharfe in its **middle course**. The river is different from the upper course, because it has:

- a gentler gradient
- a greater **discharge** (volume of water) because tributaries (like Buckden Beck) join it
- a smoother channel, because smaller pebbles, muds and sands have replaced stones and boulders
- faster **velocity** (speed) because there is less friction to slow down the river
- more energy to erode sideways (laterally).

The valley shape also changes because:

wide flat flood plain — *steep valley sides* — *meander* — *point bar*

Figure 1 *The valley of the Wharfe in its middle course. This is a meander and flood plain near Kettlewell in Wharfedale.*

- the **flood plain** widens the valley floor (the valley is now a U-shape)
- deposits of sands and clays (**alluvium**) fill the flood plain, and are fertile for farming
- the river now winds or **meanders**, creating **ox-bow lakes** (Figures 1 and 2).

How meanders change the valley

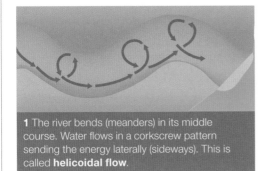

1 The river bends (meanders) in its middle course. Water flows in a corkscrew pattern sending the energy laterally (sideways). This is called **helicoidal flow**.

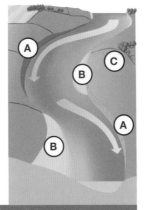

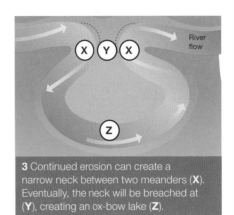

3 Continued erosion can create a narrow neck between two meanders (**X**). Eventually, the neck will be breached at (**Y**), creating an ox-bow lake (**Z**).

2 The fastest current (the **thalweg**) undercuts the bank on the outer bend (**A**). This produces a steep edge (**river cliff**), which eventually collapses, so the channel moves. Sediment is deposited on the inner bank (**B**) by slower moving water, to form a point bar, and eventually a flood plain (**C**).

Figure 2 *The formation of meanders and ox-bow lakes*

Six Second Summary

- Increased discharge and velocity in a river's middle course provide more energy, and it erodes laterally.
- In the middle course, valleys are U-shaped with a flat valley floor.
- Flood plains covered in alluvium widen the valley floor.
- Meanders are natural bends in the river.

Over to you

Draw a table to compare **a)** the river and **b)** the valley shape, in the middle course with the upper course.

*Student Book
See pages 146–147*

The lower course of the Wharfe

The river is:

• wide and deep, with a gentle gradient
• has a very large discharge
• floods easily and has a wide and almost flat flood plain
• has large, winding meanders
• is affected by tides from the sea.

The Wharfe joins the River Ouse, and then the Humber to form an **estuary** where it meets the sea. Here, the river is tidal, so:

• flow is **outwards** at low tide (to the sea) or **inwards** (from the sea inland) at high tide
• at high tide, incoming flow meets outgoing flow and sediment is deposited forming **mudflats** (Figure 1).

When mudflats extend beyond the coastline, a **delta** forms, e.g. the Wash in East Anglia.

• **Salt marshes** form when the sea submerges the estuary at high tide. They are important wildlife habitats but are threatened by industry and port activity.
• **Levées** form. These are embankments formed by deposition at the side of the river (Figure 2). They can prevent flooding.

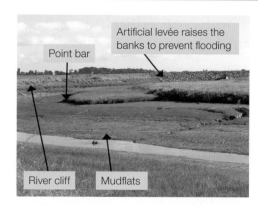

Figure 1 Mudflats near the Humber estuary

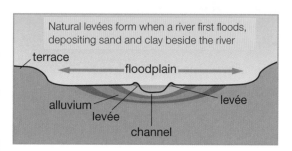

Figure 2 The features of a flood plain

How river courses change

From the upper to the lower courses of a river, there are changes in:

• river discharge and velocity
• channel width and depth
• sediment size and volume
• channel bed roughness
• **long** and **cross profiles** (Figures 3 and 4).

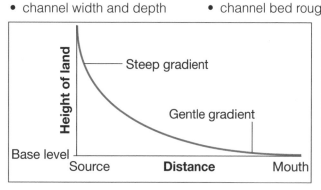

Figure 3 The long profile of a river

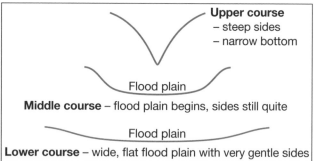

Figure 4 The cross profile of a river

Over to you

Draw a table to show the differences between **a)** rivers, and **b)** river valleys in their upper and lower course.

Geographical skills: investigating rivers and their valleys

You need to know:

- how to interpret diagrams and maps of river valleys
- how to draw a map cross section.

Identifying river valleys and landforms

Looking at a map and visualising what the landscape looks like (Figure 1) is a key skill you need to develop.

Being able to draw a cross section from a map helps to start this process.

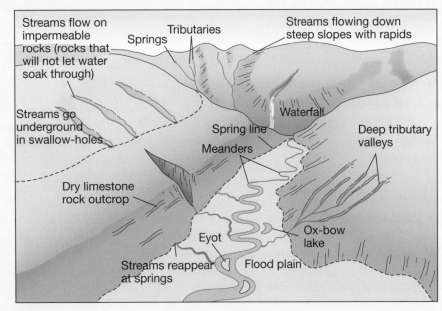

Figure 1 *Landforms of a typical upland region*

How to draw a map cross section

Stage 1

- Place a piece of paper wide enough to cover the length of the cross section on the map.
- Each time your piece of paper crosses a contour line, mark it with a pencil AND write down the height in metres. You can also mark on other features that the paper edge crosses (e.g. rivers, roads).

Stage 2

- Draw graph axes as wide as the cross section, and as high as the height of land needs (choose a suitable scale).
- Lay your paper on the x-axis (horizontal) and plot points on your graph.
- Join the points with a line which shows the relief of the landscape along the length of the cross section. How accurate that is depends on the scale of the y-axis (vertical).

Stage 1

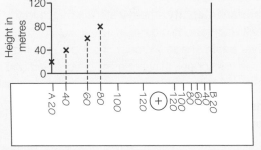

Stage 2

Figure 2 *How to draw a cross-section*

Six Second Summary

- Interpreting diagrams, sketches and maps is a key skill.
- A cross section of a map shows the profile of the area.

Over to you

Look at Figure 1. Sketch a contour map based on this diagram. Start by sketching the river and build up the contours on either side to show as many of the features as possible.

Student Book
See pages 150–151

You need to know:

• how human and physical factors affect storm hydrographs.

Storm hydrographs

A storm hydrograph (Figure 1) is a graph showing how a river reacts to a rainfall event. It shows two things:

• rain (a bar chart)
• discharge (a line graph) rises from baseflow to peak discharge (the rising limb) and then falls (the falling limb) as water from the rain storm flows away.

Human and physical factors affect the shape of a hydrograph.

Human factors

• **Land use change** – replacing fields or woodland with buildings and roads, or pasture with arable (crops) increases runoff. Rain reaches a river more quickly, reducing lag time.
• **Deforestation** reduces interception and infiltration, so the hydrograph shape resembles 'A' (Figure 1).

Physical factors

Figure 2 shows what happens when rain falls.

• Rainfall is intercepted by vegetation.
• Some of this is evaporated, the rest infiltrates the soil or flows overground as surface runoff.

How quickly **surface runoff** happens depends on:

• **antecedent** (previous) **rainfall** – recent heavy rain may saturate soil, making flooding more likely
• **permeability** – permeable rocks absorb water, so runoff is rare, but occurs quickly on impermeable rocks
• **rainfall intensity** – heavy storms cause low infiltration and rapid runoff
• **river basin shape** – circular shaped basins increase flood risk; longer, thinner basins have lower flood risk.

Once water infiltrates into the soil, it is either **transpired** by plants or reaches the river via **throughflow** or **groundwater flow** (Figure 2).

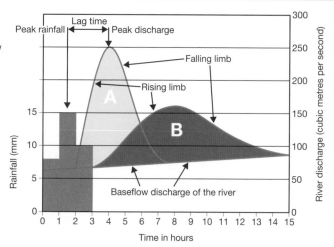

Hydrograph A Urban areas – little infiltration, rapid runoff and transfer of water to rivers. River levels rise quickly.

Hydrograph B In rural areas water is absorbed, or in summer water evaporates. Both increase the time water takes to reach a river; lag time is longer; peak discharge is lower.

Figure 1 *Hydrograph **A** shows a quick response to a rain storm; Hydrograph **B** shows a slow response*

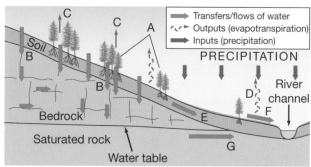

Figure 2 *Cross section of a river valley showing how water reaches a river*

A Intercepted by leaves and branches **(Interception zone)**
B **Infiltrates** (soaks) into the soil
C **Transpires** back to the atmosphere through leaves
D **Evaporates** back to the atmosphere
E Seeps through soil **(throughflow)**
F Flows over the ground as **surface runoff**
G Seeps through solid rock **(groundwater flow)**

Six Second Summary

• Rainfall is intercepted by vegetation, then infiltrates the soil or becomes surface runoff.
• Infiltrated water reaches the river through the soil (throughflow), or rock (groundwater flow).
• Storm hydrograph shape varies, depending on different factors (e.g. human activity).

Over to you

Draw a spider diagram to summarise the factors that affect the shape of a storm hydrograph. Split them into human and natural factors.

Student Book
See pages 152–153

You need to know:

- how physical and human factors caused the Sheffield floods of 2007.

Physical causes of the floods

- In June 2007 heavy prolonged (antecedent) rainfall fell across South Yorkshire. Almost 190 mm of rain fell on two days – 15 and 25 June (Figure 1). The saturated ground produced a hydrograph similar to **A** in Section 4.20.
- Sheffield is surrounded by steep hills. Surface runoff occurred quickly.
- The city lies at the confluence of the rivers Rivelin and Loxley (X on Figure 2) and Don (Y). Each river was filled to capacity so flooding was inevitable where they met.
- Reservoirs in the upper courses of each river overflowed.

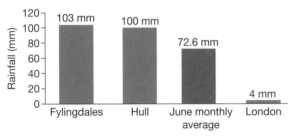

Figure 1 *Rainfall in Yorkshire on 25 June 2007 compared with the average for June, and London*

⏱ Six Second Summary

- The 2007 Sheffield floods were the result of physical factors (prolonged rain, the confluence of rivers and steepness of the landscape).
- Human factors contributed to the floods, (urbanisation, and drains being overwhelmed).
- The floods had impacts on people, businesses and transport.

✎ Over to you

- List two human causes and two physical causes of the 2007 Sheffield floods.
- Classify the impacts of the floods into economic, social and environmental.

Human causes of the floods

- Most of Sheffield's surfaces are impermeable – concrete, brick and tarmac.
- Flood prediction was difficult due to sudden downpours.
- The drainage network was overwhelmed, and in some cases blocked.
- The drainage system was designed to deal with rainfall amounts that might occur once in 30 years. But the floods were a 1 in 400 year event, so planning could not have prevented flooding.

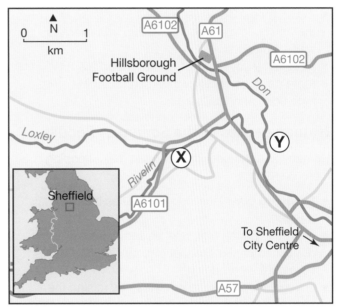

Figure 2 *The area around Hillsborough, north-west Sheffield, where some of the worst flooding occurred*

Impacts of the floods

- Two people drowned.
- Over 1200 homes flooded; 1000 businesses affected.
- Roads were damaged; rail links flooded.
- 13 000 people without power for two days.
- Hillsborough Football Stadium flooded and cost several million pounds to repair.
- Health risks as raw sewage escaped into floodwater.
- Meadowhall Shopping Centre was closed for a week, despite defences to protect it against a 1 in 100 year flood.

The Somerset Levels

The Somerset Levels are a low-lying landscape of wetlands in south-west England drained by seven rivers (Figure 1). Floods are common during spring tides in the Bristol Channel – but winter 2014 was one of the worst (Figure 2).

Physical causes of the Somerset floods

The floods resulted from high level winds (**Jet streams**) bringing low pressure weather systems across the Atlantic to southern England, which caused:

• the stormiest UK winter for over 20 years
• the most rainfall since 1766 – 235% of average winter amounts!
• more rainy days than any time since 1961
• more days when high winds combined with very high tides and tidal surges.

Human causes of the Somerset floods

Human interference with rivers can make flooding more likely.

• **Dredging** creates levées and makes the channel deeper.
• However, during high rainfall, sediment is deposited on the channel bed, raising it to where it was before dredging!
• Levées make the problem worse by raising the river bed further, as the river naturally adjusts to the raised banks.

The future

Climate scientists are clear that climate change will bring the following for the UK:

• more storms and damaging winds   • higher spring tides
• higher, longer-lasting floods   • more storm surges.

Key
1 Tone **2** Yeo **3** Parrett
4 Axe **5** Brue **6** Huntspill
7 King's Sedgemoor Drain

Figure 1 The Somerset Levels; flooded areas are in blue

Figure 2 The village of Moorland surrounded by floodwater in winter 2014

Student Book
See pages 156–157

The options ahead

There are debates about how to manage flood risks on rivers. Flood protection is carried out by the Environment Agency which chooses between hard and soft-engineering methods.

Hard-engineering methods

These include structures (e.g. flood walls, flood relief channels, levées) and ways of managing the channel (e.g. dredging) to defend areas from floodwater.

Method	How it works	How effective is it?	Cost per km
Flood walls	Increases river's capacity to prevent flooding.	• Cheap, 'one-off' cost – once built, it's done. • Useful in city centres, where space is limited. • Water dispersed quickly, increasing flood risk downstream.	Depends on type of wall and material used.
Construct levées	Usually built some distance from the river, increasing capacity.	• Expensive, but reduced fear of flooding for those close to river. • Can increase the flood risk downstream. • Can fail by overtopping (water rises over the levée), slumping, or by erosion.	Up to £1m depending on materials used.
Dredging	Increases channel capacity. Lining with concrete speeds up river flow to get flood water away quickly.	• Needs to be done often as the channel fills with sediment each time it rains heavily. • Concrete lining is expensive but cheap to maintain. • Speeding up flow increases the flood risk downstream.	£50 000
Flood relief channel	Create extra channels to divert excess water from city centre (e.g. in Rotherham and Exeter).	• Protects built-up areas but could cause flooding elsewhere.	• £14m/km in Rotherham 2008. • £4.3m/km in Exeter in 2015.

Figure 1 *Summary of hard-engineering methods used in flood protection*

Soft-engineering methods

These include solutions to adapt to flood risks and allow natural processes to manage rainwater. The Environment Agency now believes that:

* upstream, upland areas should be **planted with trees** to reduce surface run-off

* flood plains should be lowered or levées removed to encourage flooding over certain areas (called **flood plain retention**)
* river channels should be **restored** to their natural state (called **river channel restoration**)
* planning permission should **avoid building** near rivers.

Method	How it works	How effective is it?	Cost per km
Flood plain retention	• Flood plain is lowered, and surface restored to shrubs or grassland, so it retains water and releases it slowly into the river.	• Increased ability to store floodwater. • Reduced flooding in 2007 despite heavy rains. • The only flooding in Darlington in 2007 was caused by backlogged drains.	£1.2 million in total for a 2 km stretch.
River channel restoration	• Some meanders were rebuilt, slowing water down. • Banks were lowered so the park was flooded instead of Darlington. • Hard engineering materials were replaced with sediment, and planted with trees.	• Improved ecology with a 30% increase in birds and insects, within one year. • In a survey, 82% of people liked the more natural look 'mostly' or 'strongly'.	

Figure 2 *Summary of soft engineering methods used along the River Skerne in Darlington. The Environment Agency hopes to reduce the River Skerne's response so that it is more like hydrograph B (Section 4.20).*

 Six Second Summary

* Hard engineering includes building structures to prevent flooding.
* Soft engineering involves allowing natural processes to deal with rainwater.

 **Over to you**

Draw a spider diagram to show how different methods of hard engineering and soft engineering work.

Topic 5
The UK's evolving human landscape

Your exam

- Topic 5 The UK's evolving human landscape makes up Section B in Paper 2, UK geographical issues.

- Paper 2 is a 90-minute written exam and makes up 37.5% of your final grade. The whole paper carries 94 marks (including 4 marks for SPaG) – questions on Topic 5 will carry 30 marks.

- You must answer all parts of Section B. Section A contains questions on The UK's evolving physical landscape (pages 64–87), and Section C contains questions on fieldwork, where you have a choice of questions (pages 109–129).

Tick these boxes to build a record of your revision

Your revision checklist

Spec Key Idea	Detailed content that you should know	1	2	3
5.1 Population, economic activities and settlements are key elements of the human landscape	• Differences between urban and rural areas, and government policies which have attempted to reduce them			
5.2 The UK economy and society are increasingly linked and shaped by the wider world	• The effects of national and international migration on UK population, ethnic and cultural diversity			
	• Changes to UK economic sectors in urban and rural areas and their effect on contrasting regions of the UK			
	• The UK economy – globalisation, trade policies, privatisation, FDI and TNCs			
5.3 The context of the UK city influences its functions and structure	• Site, situation and connectivity of the city in a national, regional and global context			
	• The city's structure, functions and variations in building age and density, land use and environmental quality			
5.4 The UK city changes through employment, services and the movement of people	• The influence of national and international migration on the city			
	• Reasons for different levels of inequality in the different parts of the city			
5.5 The changing city creates challenges and opportunities	• Decline in some parts of the city (de-centralisation, e-commerce, transport)			
	• Economic and population growth in other parts of the city			
5.6 Ways of life in the city can be improved by different strategies	• Positive and negative impacts of regeneration and rebranding on people			
	• Making urban living more sustainable and improving quality of life			
5.7 The city is interdependent with rural areas, leading to changes in rural areas	• Economic, social and environmental costs and benefits of interdependence between the city and rural areas			
	• Economic and social changes in one rural area			
5.8 The changing rural area creates challenges and opportunities	• Issues of housing, employment, healthcare, education for rural people			
	• Rural diversification and tourism, and their environmental impacts			

You need to know:

- about population density of the UK, and its urban core regions.

Student Book
See pages 160–161

Our overcrowded islands?

The **population density** map (Figure 1) shows the uneven distribution of the UK's population. Urban areas account for just 7% of the total UK area – so 93% is not crowded!

But urban areas are important for the economy – London produces 22% of the UK's annual GDP with only 13% of the population. These are the **core regions** of the UK.

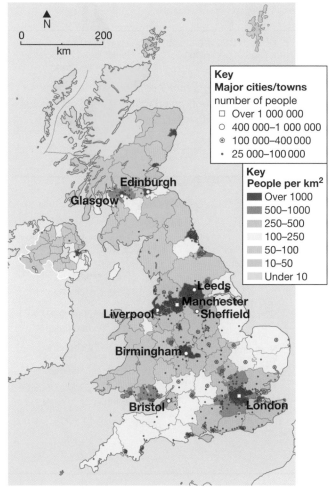

Figure 1 UK population density, 2004

Big Idea

The UK's urban areas are the driving force behind the economy.

The UK's urban core

The UK's densely populated cities also attract people for work.

- They spend money on housing, goods and services, creating jobs – the **multiplier effect**.
- As the multiplier effect develops, the city becomes the centre of a core region.
- Cities can merge with towns to form **conurbations**, which influence an even larger area – e.g. the region from which people commute to work.
- They spend their money locally, boosting the economy of a wider region.

Economic activity

Outside London, the UK's conurbations initially grew around Britain's coalfields as:

- mining, metalworking and engineering cities (e.g. Birmingham); Newcastle and Glasgow were shipbuilding centres
- textile manufacturing centres (e.g. Manchester).

Now, manufacturing has been replaced by service industries (e.g. finance, property development). Liverpool, Manchester, Leeds, Bradford and Sheffield form the '**northern powerhouse**' – a core region with potential to drive the economy of northern England.

 Six Second Summary

- Population density varies throughout the UK.
- Urban core regions attract people, and the multiplier effect ensures their growth.
- Old manufacturing cities have developed service industries, and northern cities have become a major core region.

 Over to you

Look at Figure 1.

- State three things about the UK's population distribution.
- Why is there a small area of dense population on the east coast of Scotland?

You need to know:

*Student Book
See pages 162–163*

- about the UK's rural periphery.

The rural periphery

People are often happier in rural places, where housing is cheaper and takes up less income (the **rural periphery**). They have:

- **a lower population density** with smaller towns and villages, and isolated farms.
- **older populations** (Figure 1) because they're popular for retirement – cities often attract younger people with a busier lifestyle.
- **lower incomes** due to older populations with pensions, and seasonal/part-time jobs with lower wages in farming and tourism.
- **high transport costs** – a car is often essential due to lack of public transport and long commutes.
- **out-migration of younger people** to areas with opportunities for university or employment.

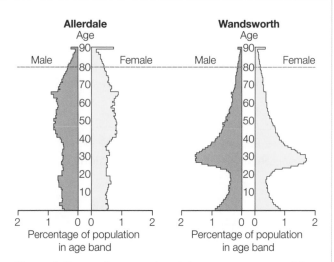

Figure 1 *Comparing a rural periphery age-sex pyramid (Allerdale in the Lake District) with that of an urban area (Wandsworth in south-west London)*

Reducing the gap

Rural peripheral areas don't receive the same levels of investment as **affluent** urban-core areas. The government and the EU have four incentives for businesses and investors:

- **Enterprise zones** (Figure 2) – help is given with start-up costs, reduced taxes, and access to superfast broadband.
- **Regional development grants** to help start businesses in peripheral areas.
- **EU grants** – for regions with GDP below 75% of the EU average. In 2015, only Cornwall, west and north Wales qualified.
- **Improvements to transport** – most investment is in England's urban core, e.g. the HS2 railway. Cuts in government budgets reduced spending on transport in rural areas (except in Scotland).

Figure 2 *Enterprise zones in the UK in 2015; most are in urban areas and all are in England*

Six Second Summary

- The rural periphery has cheaper housing and people are happier there.
- The rural periphery has older populations, lower incomes, high transport costs and out-migration of young people.
- Measures have been taken by the UK government and the EU to help poorer regions.

Over to you

Summarise four incentives for investment in rural areas that are available in the UK.

*Student Book
See pages 164–165*

You need to know:

- how the UK's population geography is changing.

A fast-changing country

The UK's population grew from 54.3 million in 1965 to 65 million in 2015. There were two causes: **net immigration** and a **rising birth rate**.

Net immigration

In 2014 net migration in the UK (the difference in numbers of emigrants and immigrants) was 318 000. Immigration was due to:

- EU membership – workers moving to the UK
- **globalisation** – highly qualified and skilled workers moved to London's knowledge economy.

A rising birth rate

The UK's rising birth rate has been caused by:

- women choosing to have children earlier (due to reduced employment after the 2008 recession)
- more older women choosing to have children
- increasing numbers of overseas-born women who often have higher fertility rates than UK – born women (e.g. for religious reasons).

The impacts of immigration on the population

The UK population is now more **multicultural** than at any time in its history.

- 37% of London's population was born overseas.
- The UK has an ageing population, so needs more working people to pay tax (to help with pensions and healthcare).
- Immigration brings social benefits (e.g. cultural diversity).

Changing population distribution

Nearly 3 million people a year move from one area of the UK to another. Add this to net increase in population (births, deaths, migration), and there's a difference in population growth across the UK (Figure 1).

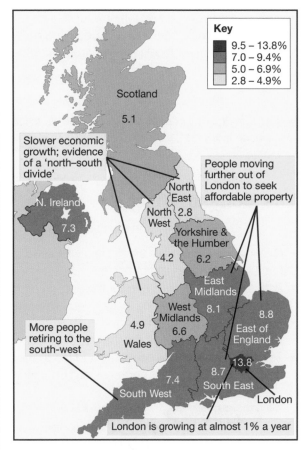

Figure 1 *Percentage increases in UK population using census estimates, 2003–13.*

 Six Second Summary

- The UK's population is growing due to net immigration and a rising birth rate.
- Net immigration has increased due to EU membership and globalisation.
- The UK has an ageing population.
- Population in some areas is growing at a faster rate than others.

Over to you

Give two reasons why the UK's population is growing so quickly.

You need to know:

- about changes in the UK's primary and secondary economic sectors
- the reasons for these changes.

*Student Book
See pages 166–167*

From the old economy to the new

Until 1992, Dinnington in South Yorkshire had one of the largest coal mines in Yorkshire providing **primary sector**, well-paid employment.

- The colliery closed in 1992 and is now the site of a business park for mainly **tertiary** services.
- This change from primary/secondary employment to tertiary is common throughout the UK (Figure 1).
- Jobs available in Dinnington now tend to be temporary and/or part-time and poorly paid.
- There are also few local jobs so a commute to cities like Sheffield is necessary.

Figure 1 *Changes in UK employment, from the old to the new economy, between 1980 and 2015 (millions of workers)*

Date	Primary	Secondary	Tertiary and Quaternary
1980	0.89	8.9	17.6
1985	0.78	7.3	18.3
1990	0.68	7.5	21.0
1995	0.54	6.3	21.0
2000	0.48	6.2	22.9
2005	0.46	5.6	25.2
2010	0.46	5.0	26.0
2015	0.48	5.1	28.1
Trend	**Down by 45%**	**Down by 43%**	**Up by 59%**

Jobs almost halved; mining lost almost 250 000 jobs and farming lost 100 000 jobs in this period.	Employment also fell sharply; primary/secondary sectors form the traditional **old economy**.	This sector has expanded and become the basis for the **new economy**; tourism replacing farming in rural areas.

The 'domino' effect

Changes to one industry can affect another.

- Coal was used to make steel; steel was used in engineering, manufacturing and shipbuilding.
- Both suffered from overseas competition, creating a domino effect – the collapse of one industry led to the collapse of another.
- Shipbuilding on the Tyne ended in 2007, the last deep coal mine closed in 2015, as did the Redcar steel works.
- This affected the local economy and led to the **de-industrialisation** of north-eastern England (Figure 2).

Figure 2 *The 'dominoes' – coal, steel, engineering and shipbuilding*

What's left in the North?

Although manufacturing has declined, Nissan in Sunderland is Europe's largest car factory, and Sheffield produces as much steel as ever. However, most employment is now tertiary, including:

- transport companies DB Arriva and Go-Ahead
- call centres, e.g. Tesco (wage rates are 40% lower than London)
- there is a lack of high-salary jobs. Leeds is an important centre for financial services, but is small compared to London.

Six Second Summary

- UK primary and secondary industry has declined rapidly and the tertiary sector has grown, changing the employment structure.
- As one industry collapses, it impacts on other industries and local businesses/services.

Over to you

Draw a spider diagram to show how the UK's employment structure has changed from the old to the new economy.

*Student Book
See pages 168–169*

You need to know:

- what is meant by tertiary and quaternary sectors
- how and why these have expanded in the UK.

The knowledge economy

100 000 people work in Canary Wharf in London's Docklands.

- Many work long days in banking and investment trading with the East (seven hours ahead of UK time), and also New York (five hours behind).
- Average salaries of £100 000 a year (four times higher than the UK average) make the long days worthwhile.

Faced with declining primary and secondary employment in the 1980s, the UK government encouraged **tertiary and quaternary** sectors to develop a '**new economy**' (Figure 2).

- Both sectors offer services, but quaternary offers highly specialised services and skills – the **knowledge economy**.
- University degrees, plus specialised training, are essential.

The knowledge economy in Canary Wharf includes companies working in UK and overseas property development, law, insurance, IT, and the creative industries.

The new rural economy

The knowledge economy is not restricted to cities.

- Companies in the new economy are **footloose** – they're not tied to location.
- Working at home (**teleworking** – using the Internet and broadband) is popular and avoids companies renting expensive offices.

Advantages	Disadvantages
Better health; people take breaks during the day	Isolation from work colleagues and less contact with your boss; may lead to being overlooked for promotion.
No commuting – reduced stress and longer hours worked	
Less absenteeism and sickness	It's sometimes difficult to motivate and organise home-workers
Parents save money on childcare	
May suit disabled people who don't have to travel	Work never disappears – it is always around you and difficult to 'switch off'
Allows variable hours of work	

Figure 1 *Advantages and disadvantages of teleworking*

Low salary 'new economy', e.g. Dinnington and retail centres	High salary 'knowledge economy', e.g. London's Canary Wharf
Sector: Tertiary	**Sector:** Quaternary
Examples: Jobs with delivery firms, in retail parks, or shopping centres. Jobs advertised locally.	**Examples:** Jobs in global banking or law. Jobs advertised globally to get the right people.
Located: On the outskirts of towns for cheaper land and local labour.	**Located:** Where there are highly skilled, educated staff, good IT and broadband.
Qualifications: Mostly unskilled, needing few qualifications.	**Qualifications:** A degree and training, e.g. law, finance, IT.
Part-time or full-time?: A quarter of jobs are part-time; many are temporary, lasting weeks/months (e.g. Christmas).	**Part-time or full-time?:** Mostly full-time; contract jobs also available (with high daily rates of pay).
Earnings: Wages usually low (minimum wage or just above), with more for supervisor roles (e.g. in supermarkets).	**Earnings:** High salaries plus bonuses. Salaries depend on qualifications; many earn six-figure amounts.
Employee mix: Mix of male and female, with women the majority.	**Employee mix:** Mostly male, especially in banks or property companies.

Figure 2 *Summary of 'new economy' and 'knowledge economy'*

Six Second Summary

- The quaternary sector (knowledge economy) is based on specialised knowledge and skills.
- Technology allows those in rural areas to work in the knowledge economy.
- The new economy is a service-sector economy.
- Many jobs in the new economy are poorly paid; those in the knowledge economy are much better paid.

Over to you

List the advantages and disadvantages of **a)** working in the tertiary sector compared to the primary or secondary sectors, **b)** working in the knowledge economy.

Globalisation

Virgin shows how global many UK companies have become.

- From a small beginning in entertainment in 1968, Virgin now consists of over 400 companies, including travel, financial services, healthcare, food, drink, and telecommunications.
- It operates in over 50 countries with over 50 000 employees.

TNCs and FDI

The UK's economy has become increasingly globalised since the 1980s. TNCs like Virgin expect:

- the free flow of goods and services (**free trade**), without tariffs (see Section 2.5). Most TNCs supported EU membership because of its free trade.
- to **employ people** wherever it is cheapest. The EU allows access to a huge labour supply.
- to invest anywhere with unrestricted flows of capital, known as **Foreign Direct Investment** (FDI – see Section 2.7).

With FDI, UK companies can invest abroad, but overseas TNCs also invest in the UK.

- In 2014, the UK received more FDI than any other country in Europe.
- If your family has ever bought or rented a Jaguar or Land Rover, drunk Tetley Tea or even driven on roads treated with salt then you've been helping Tata, an Indian TNC, to earn money.

TNCs and privatisation

Privatisation is the change in ownership of services from the public sector (run by national or local government) to the private sector (owned by shareholders). Some think it leads to increased efficiency, others think that profit should not be made from basic needs.

In the UK, privatisation has taken place in:

- **infrastructure** – many UK energy, water and rail companies are overseas-owned
- **local councils** – some services are contracted to private companies, e.g. Aeolia or Serco
- **the NHS** – many services are contracted to private companies, e.g. Virgin Care (Figure 1).

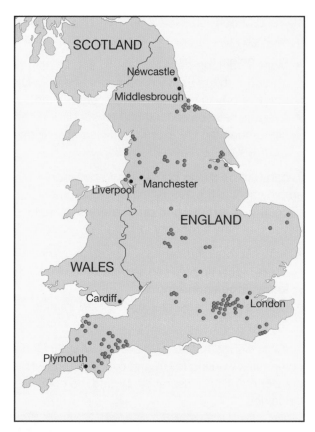

Figure 1 *The location of Virgin health and care services in the UK, 2015*

 Six Second Summary

- The UK's economy has become increasingly globalised since the 1980s as a result of FDI and free trade.
- FDI allows UK companies to invest abroad, and overseas TNCs to invest in the UK.
- In the UK, privatisation has increased the role of TNCs.

 Over to you

Close your book. Define these terms:
a) globalisation, **b)** TNCs, **c)** free trade, **d)** FDI,
e) privatisation.

Student Book
See pages 172–173

You need to know:

- why London's site and situation is important.

London's site

In 43 AD, the Romans bridged the Thames near the present-day London Bridge.

- The **site** chosen was the last place where the Thames was shallow enough to cross before reaching its estuary.
- The bridge attracted market traders, and within decades a significant town had grown, Londinium, which became the capital by 200 AD.
- By 1300 further growth brought sea traders and people searching for work.

Big Idea

The original site chosen for London was crucial for its future success as a modern world city.

London's situation

An important factor in London's growth is its **situation**, i.e. its location within the UK.

- As a port, London could trade by sea quickly with nearby Europe.
- Even when the industrial cities of the Midlands and North were growing during the Industrial Revolution, London had a bigger population, economy and port.
- Internationally, London's time zone helps its economic growth today. Those working in London's finance industries (Section 5.5) can trade with Asia (5–7 hours ahead), Australia (8–11 hours ahead), and later in the same day with New York (5 hours behind).

London's cultural diversity

The different groups that have migrated to London for centuries have added to its cultural diversity. Now London's knowledge economy makes it a magnet for migrants, and its schools teach students from over 200 countries.

The wider world

London has managed to keep its position as a major 'world city'. Its **connectivity** (how easy it is to travel or connect with other places) has been a major factor:

- **Internationally**: London's airports make it the world's largest international air 'hub'. Eurostar brings European cities within a few hours of London.
- **Nationally**: the fastest rail services link London and major UK cities. HS2 will reduce journey times further. Urban core regions are brought closer to London, but peripheral areas seem further away (Figure 1).
- **Regionally**: Major roads link London with other major cities.

Figure 1 Map of the UK distorted to show fastest travel time by rail to London, rather than physical distance

Six Second Summary

- London's site and situation have been important factors in its growth.
- Today, London is a 'world city' in terms of its economy, connectivity and its cultural diversity.

Over to you

Create a spider diagram to explain how London's connectivity has helped it remain a 'world city'.

Student Book
See pages 174–175

You need to know:

- about London's structure and functions, and how it varies.

Chaos or order?

From ground level, cities may seem chaotic. But Figure 1 shows a structure with the high-rise offices of the **Central Business District (CBD)** clearly visible.

London's CBD

The CBD is usually the **oldest** part of a city, with most of its offices.

- London's radial road network means the CBD is accessible which increases land values (Figure 2).
- The CBD is **densely built**, with the highest buildings (because land is expensive).
- Despite the density, central London's environmental quality benefits from its royal parks (e.g. Hyde Park).
- But London also has the UK's worst air quality, caused by road traffic.

London's inner suburbs

During the Industrial Revolution (18th/19th centuries), factories and high-density terraced housing were built close to central London. A few high-income suburbs developed close to the city. These inner suburbs are London's most varied:

- 1 km west of the West End is Kensington – one of the world's most expensive suburbs.
- 1 km east of the City is Hackney – an area of older factories which are being replaced with newer flats.

The inner suburbs are changing rapidly.

- Large older houses have been divided into rented flats.
- Environmental quality varies between areas that are run-down but changing (e.g. Hackney) and smarter areas like Notting Hill near London's parks.

The Shard – a recent expansion of the CBD around London Bridge railway station

The City of London – the CBD consisting of banks and offices

Figure 1 Central London's high-rise offices that make up the CBD; with Canary Wharf and the 'West End', London actually has three CBDs!

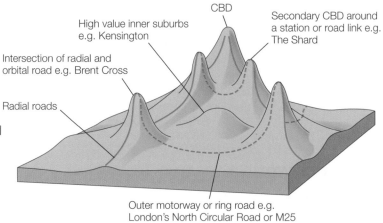

CBD

High value inner suburbs e.g. Kensington

Secondary CBD around a station or road link e.g. The Shard

Intersection of radial and orbital road e.g. Brent Cross

Radial roads

Outer motorway or ring road e.g. London's North Circular Road or M25

Figure 2 A general pattern of urban land values. The CBD is the highest value part of the city; other high-value areas occur where roads intersect (or meet).

London's urban–rural fringe

Loughton is on the edge of Epping Forest on London's **urban–rural fringe**.

- Most houses were built in the late 20th century and have gardens, so density is lower.
- There is some industry.
- Environmental quality is higher.

Six Second Summary

- Cities have a structure with the CBD at the centre (the oldest part).
- Outside the CBD are the inner suburbs, which are changing rapidly.
- The urban–rural fringe contrasts with inner city areas.

Over to you

Look at Figure 2. Sketch a similar diagram and replace the names with those from your local area or a city that you know.

Student Book
See pages 176–177

You need to know:

- why people migrate to London
- how migration affects the character of different parts of London.

London's changing population

Internal and overseas **migration** (Section 5.3) caused London's population to grow faster in 2012 than at any time in its history. 1.9 million **overseas migrants** settled in London between 2000 and 2013.

Who are the migrants?

Internal migrants are mainly UK graduates seeking work and a London lifestyle. International migrants consist of **skilled** and **unskilled** workers.

- Many **skilled workers** take well-paid jobs in the knowledge economy in the City (Section 5.5).
- Most tend to be white and highly qualified from the EU, USA, South Africa and Australia.

- **Unskilled workers** do jobs not wanted by UK workers or with unsocial hours (e.g. refuse collection, pizza delivery).
- Many come from the EU, India, Pakistan, Bangladesh and West Africa.

Migration and ethnic communities

Recent migrants are not eligible for social housing, so seek cheap, rented accommodation. Often, clusters of ethnic communities develop in inner city areas. These:

- help to defend migrants against discrimination
- support ethnic shops and services (banks, places of worship)
- help to preserve cultural distinctiveness, e.g. festivals such as the Notting Hill Carnival.

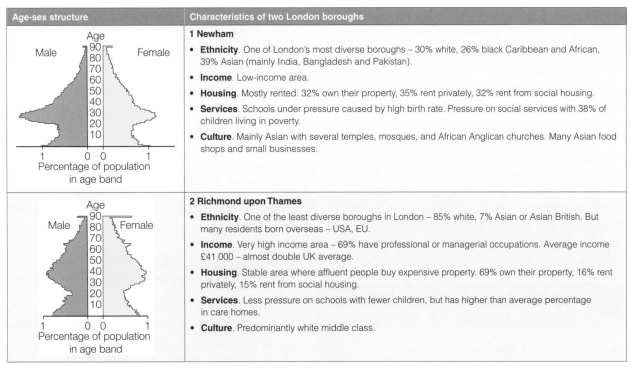

Age-sex structure	Characteristics of two London boroughs
	1 Newham • **Ethnicity**. One of London's most diverse boroughs – 30% white, 26% black Caribbean and African, 39% Asian (mainly India, Bangladesh and Pakistan). • **Income**. Low-income area. • **Housing**. Mostly rented. 32% own their property, 35% rent privately, 32% rent from social housing. • **Services**. Schools under pressure caused by high birth rate. Pressure on social services with 38% of children living in poverty. • **Culture**. Mainly Asian with several temples, mosques, and African Anglican churches. Many Asian food shops and small businesses.
	2 Richmond upon Thames • **Ethnicity**. One of the least diverse boroughs in London – 85% white, 7% Asian or Asian British. But many residents born overseas – USA, EU. • **Income**. Very high income area – 69% have professional or managerial occupations. Average income £41 000 – almost double UK average. • **Housing**. Stable area where affluent people buy expensive property. 69% own their property, 16% rent privately, 15% rent from social housing. • **Services**. Less pressure on schools with fewer children, but has higher than average percentage in care homes. • **Culture**. Predominantly white middle class.

Figure 1 *The effect of migration on two contrasting London suburbs*

 Six Second Summary

- London's population is changing due to internal and overseas migration.
- Migrants include both skilled and unskilled workers.
- Some ethnic groups cluster in specific areas.
- Migration has affected areas culturally.

Over to you

Draw a spider diagram to illustrate the different types of migrants in London and the reasons why they migrate there.

You need to know:

Student Book
See pages 178–179

- why there are inequalities in different parts of London.

London's inequalities

In 2012, 28% of London's population lived in poverty (7% higher than the rest of England).

- One million of the UK's poorest people, and one million of its wealthiest, live in London.
- Incomes there are more unequal than any other part of the UK.
- There is a close link between deprivation and life expectancy.

Comparing Newham and Richmond upon Thames

Newham in east London is one of London's most deprived boroughs, while Richmond in south-west London is one of its wealthiest (Figure 1). Figure 2 shows that deprivation and health are closely linked.

- Incomes are low in Newham, so the borough has greater numbers of children on free school meals.
- Average household incomes in Richmond are twice those of Newham.
- The higher percentage of those in Richmond with degrees enables them to gain better-paid employment.
- Health in Newham is worse than in Richmond. Ill health can limit people's ability to work.
- In both areas, the percentage of 19-year-olds without qualifications is high, restricting them to low-paid unskilled work.
- Newham almost equals Richmond for the percentage of students gaining five GCSEs at grade C/4 or above because its schools have improved.

	Newham	Richmond
General health		
Infant mortality (per 1000 births)	5.5	2.75
People with a limiting long-term illness (%)	12.3	7.6
Premature deaths (before 65, per 100 000 population)	210	121
Education		
Percentage students age 16 who got 5 GCSEs at A*–C in 2012	62	63
% of 19-year-olds with no qualifications	41	37
% 5 to 16-year-olds taking free school meals	20	8.4
% adults educated to degree level	26	64

Figure 2 Comparing health and education indicators for Newham and Richmond, 2012

Multiple deprivation

The government gathers census data on employment, health, education, housing and services.

- These are used to produce an **Index of Multiple Deprivation (IMD)**.
- The IMD shows how deprived places are.
- Highest crime rates occur in the most deprived urban areas.

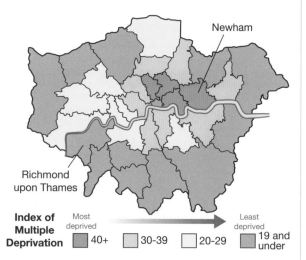

Index of Multiple Deprivation — Most deprived → Least deprived: 40+ | 30-39 | 20-29 | 19 and under

Figure 1 The Index of Multiple Deprivation for London boroughs

Six Second Summary

- The Index of Multiple Deprivation measures how deprived places are.
- Incomes, health, education and life expectancy vary across London's boroughs.

Over to you

Draw a table to compare deprivation in Newham and Richmond in terms of incomes, health and education.

Student Book
See pages 180–181

You need to know:

- the reasons why parts of London suffered decline in the past.

London's decline

London has not always been a thriving city.

- Figure 1 was typical of the riverside between Tower Bridge and the Thames estuary when London's docks closed in 1981.
- Industries that relied on the port moved or closed (**deindustrialisation**, Section 3.3).
- In 1971, 30% of London workers were in manufacturing. In 2001, it was 7.5%.
- Unemployment rates reached over 60% in parts of east London.
- From 1971–1981, over 500 000 people left inner London boroughs searching for work (**depopulation**) – 16% of its population!

Suburbanisation

Depopulation speeded up London's **suburbanisation** as people moved to the outer suburbs.

- This was made possible by improvements in the transport network
- **Electrification** of surface rail was completed in 1920, and the **underground** network established by 1930.

Decentralisation

People spent money in the suburbs rather than in London, causing **decentralisation** – shifting the balance of shopping activity and employment away from the CBD. It led to the growth of:

- **out of town shopping centres**, e.g. Bluewater in Kent close to the M25.
- **retail parks**, built away from suburban shopping centres, but close to major roads (e.g. Kew Retail Park on the South Circular Road).
- **business parks**, e.g. Stockley Park near Heathrow Airport.

Online shopping (**e-commerce**) has decentralised shopping further.

The fight back

To attract people back to London, two shopping centres have been developed at:

- Stratford in east London, accessible by surface rail and tube (Figure 2).
- Shepherd's Bush in west London, close to tube, surface rail and the M40.

Figure 1 A derelict former flour mill beside the Royal Docks

Figure 2 Stratford's Westfield shopping centre – the largest in Europe!

Six Second Summary

- The closure of London's docks had a huge impact:
  - deindustrialisation – industries closed or moved
  - depopulation – unemployment caused people to move in search of work.
- Suburbanisation led to decentralisation.
- London has attempted to attract people back to the city.

Over to you

On a mind map, summarise the reasons why parts of London declined.

Student Book
See pages 182–183

You need to know:

- how London reversed a period of declining population.

The sprawling city

Even though it lost 1.5 million people from 1951–1981, London still managed to grow in size for the reasons described in Figure 1.

Reason for growth	Effect
Counter-urbanisation	People moved to the surrounding counties, increasing their population and blurring the boundary between city and countryside.
Suburbanisation (Section 5.11)	Moving to the outer suburbs to a bigger house/garden meant the same number of people took up more space.
Falling family size	Fertility rates fell from almost 3 in 1961 to 1.6 by 2011. But when those born in the 1960s started their families, they needed more homes and space!
Divorce/later marriage	More single people, so more homes needed.

Figure 1 Reasons for London's growth from 1951–1981

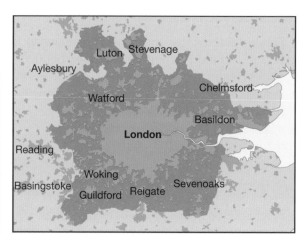

Figure 2 London's green belt

London has sprawled (it's 70 km north-south and east-west) into the countryside along the rural–urban fringe.

- To limit growth, a **green belt** (Figure 2) was created where no major building was allowed.
- Further expansion could only occur beyond it, in cities such as Chelmsford.

Re-urbanisation

Since 1991, four factors have caused the **re-urbanisation** of London.

Factors in re-urbanisation	Effect
Space	The closure of London's docks and industries (Section 5.11) created space for new housing and offices on these **brownfield sites (regeneration)**.
Investment	TNCs created jobs in financial and business services, especially Canary Wharf. HSBC has its global headquarters there.
Gentrification	Many high-income earners now prefer to live closer to work than commute. Former working-class areas in inner London (e.g. Bermondsey) have been gentrified.
Studentification	Demand from overseas students has caused university expansion, impacting on communities where students live in large numbers (studentification). Universities bring employment (e.g. lecturers), and student spending regenerates pubs, shops and buy-to-let properties.

Figure 3 Reasons for London's re-urbanisation since 1991

Culture and leisure

The 2012 Olympic Games regenerated east London.

- Before the Games, the Queen Elizabeth Olympic Park was derelict and an industrial area.
- The Olympic Stadium hosts athletics events and concerts, and is now home to West Ham United.
- A new cultural quarter is being developed.

Six Second Summary

- London is a sprawling city that is growing in size.
- The green belt limits London's size and protects the countryside.
- Re-urbanisation has occurred in London since 1991.
- The 2012 Olympics helped to regenerate east London.

 Over to you

- Learn the meaning of counter-urbanisation, suburbanisation, re-urbanisation, regeneration, gentrification and studentification.
- Write one sentence including as many of these words as possible.

You need to know:

* the impacts of regenerating and rebranding parts of London.

*Student Book
See pages 184–185*

Regeneration and rebranding

The 2012 Olympic Games were a great success in many ways.

* The site of Olympic venues had been one of Europe's biggest industrial areas,
* Its decline in the 1970s (Section 5.11) led to dereliction (Figure 1a).
* Clearing the land to create the Olympic Park was one of London's biggest projects.

East London's rising population

The Queen Elizabeth Olympic Park (Figure 1b) is one of several regeneration projects.

* They have reversed the population decline.
* Tower Hamlets lost nearly 40% of its population between 1951 and 1981 but has since grown by 58%.
* London has become a rapidly growing population of young professionals.

London suffered counter-urbanisation, but is now where many want to live.

* Its inner city suburbs have been **rebranded** (had a change of image).
* Regeneration used derelict land for housing, offices and hotels.
* Transport improvements have made east London accessible.

But housing is expensive:

* population growth is faster than the rate at which houses are being built
* overseas investors buy property but leave it vacant
* affordable housing is needed desperately.

1a 2005 (before)

1b 2015 (after)

Figure 1 *Before and after – derelict land in what became the Queen Elizabeth Olympic Park*

Changing environmental quality

Regeneration has improved environmental quality but little open space has been created.

* Queen Elizabeth Olympic Park was the first new major park in 150 years.
* Many suburbs have little open space nearby.
* Population density is increasing.

Economic opportunities

London's growing economy creates jobs, attracting people.

* About 35 000 new jobs will be created every year until 2036.
* The construction industry is booming.
* Demand for housing drives up prices.
* Most people living in London are not high earners and high rents/ mortgages make life difficult.

Six Second Summary

* Rebranding makes areas more desirable, increasing their population.
* Regeneration has improved environmental quality in parts of London.
* London's thriving economy means that it's an expensive place to live.

 Over to you

Draw a set of scales to show whether the advantages of regeneration in London outweigh the disadvantages.

Geographical skills: investigating changing environments

You need to know:

- how to use geographical skills to investigate land use in east London
- how to use 4- and 6-figure grid references and interpret an OS map.

Identifying land uses in east London

The part of east London shown on the Ordnance Survey map contains the Queen Elizabeth Olympic Park.

- The river that runs from north to south through the centre of the map is the River Lea.
- The area around the River Lea was London's most industrial area until the 1970s, when it began to decline.
- You can spot the remaining industrial areas by the larger buildings (e.g. the area between grid references 372847 and 374840).

Figure 1 *1:25 000 Ordnance Survey (OS) map extract of the Stratford area of east London. The area contains the Olympic Park, where London's 2012 Games were held.*

Big Idea

Maps can tell you about land use in an area.

Six Second Summary

- For 4- and 6-figure grid references, the first 2 or 3 numbers are the numbers from left to right, the second 2 or 3 numbers are the numbers that go from bottom to top.
- Most maps have a scale bar which shows what the real distance on the ground for any measurement taken from a map.
- Large, regular-shaped buildings on a map are usually industrial or retail developments.
- Parallel roads with high-density buildings indicate older terraced housing.

Over to you

Look at Figure 1.

1 a Give a 4-figure grid reference of an area of older terraced housing.
  b Give a 6-figure grid reference for an area of industry (shown by the letters 'Wks' – short for 'Works').
  c Give a 6-figure grid reference for the Aquatics Centre, used at the 2012 Olympic and Paralympic Games.
2 Using evidence from the map, why do you think the Lea Valley was a good site for the Olympic Games?
3 On the map, one grid square is 1 km (3.75 cm = 1 km). As the crow flies (in a straight line), how far is it from the Olympic Stadium to:
  a Stratford International Station
  b the Aquatics Centre?
  Give the compass direction for each of **a** and **b**.

You need to know:

- how London is trying to become more sustainable and improve quality of life.

*Student Book
See pages 188–189*

The daily grind

Every day, three million people travel to work (**commuting**) in London.

- Half go into central London.
- There are also large employers elsewhere, e.g. Heathrow Airport.
- London accounts for 75% of all commutes made by train in England.
- Commuting affects people's **quality of life**.

Can London become more **sustainable** (see Section 3.12) so that quality of life is improved?

A more sustainable London?

London has six related problems, which are being addressed to try to make the city more sustainable.

Transport Transport is one of the main causes of greenhouse gas emissions.

- The congestion charge, introduced in 2003, has resulted in a 6% increase in bus passengers
- Income from it is invested in public transport (£1.2 billion in ten years).
- Since 2012, all new buses have been hybrid, making them cleaner and more fuel-efficient.
- 'Source London' provides charging points for electric vehicles across London.

Affordable housing Cheaper housing lies outside London, which means longer commutes.

- The qualifying salary for affordable housing in the East Village in Stratford is £60 000 (those on low incomes are squeezed out).
- FIRST STEPS is a programme to help Londoners on a low income to buy, by offering shared ownership.

Energy efficiency Can affordable houses be energy efficient and cheap to run?

- BedZED (Beddington Zero Energy Development) in Sutton, south London, is a sustainable community of housing, offices and workplaces that promotes energy conservation. BedZED's homes use 81% less energy for heating than average, and recycle 60% of their waste.

Green space Green space is essential for quality of life. But some think housing demand in London can only be met by building on greenfield land (Section 3.6). This has disadvantages:

- Loss of farmland. The area lost to urban development nationally since 1945 is 750 000 hectares.
- Loss of rural scenery.
- Many question whether London's 'Green Belt' (Section 5.12) can survive.

Employment Can employers be persuaded that teleworking (working from home) is possible?

- Numbers of those working from home doubled from 4.3% to 8.6% in 2012.
- Flexible working hours help people to travel more cheaply outside the normal rush hour.

Waste and recycling By 2020, London aims to reduce household waste by 10%. It plans to achieve this by:

- Providing accessible recycling and composting services.
- Developing waste-burning power stations to generate heat and power.

Six Second Summary

- Commuting to work affects people's quality of life.
- London has six problems that are being addressed to increase sustainability.

Over to you

List two points on each of the six problems that affect quality of life in London.

Student Book
See pages 190–191

You need to know:

- how accessible rural areas near London are dependent on the city.

Life in Terling

Terling is a traditional-looking English village, near Chelmsford, Essex (Figure 1).

- Services are struggling – the shop is in difficulty; the doctor's surgery is only open five hours a week; the bus runs twice a week; the pub has closed.
- Dairy farming became uneconomic and so farms changed to arable (crop) farming.
- Now, outside contractors come in to plough, sow and harvest – then leave.

However, property prices are booming, and the village primary school is full.

- This is because Terling is **accessible** for London commuters (Figure 2).

Figure 1 *The church and village green at Terling, Essex*

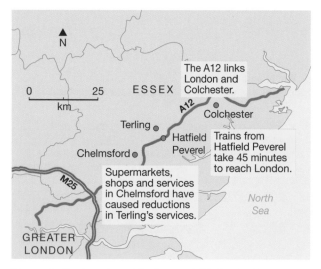

Figure 2 *The location of Terling and Chelmsford on London's rural–urban fringe*

Flows of people

The flow of people moving from London to the **rural-urban fringe** in Essex has been continuous.

- Chelmsford's population rose from 58 000 in 1971 to 168 000 in 2011, caused mainly by people migrating out of London (counter-urbanisation) for cheaper housing.
- London depends on the rural–urban fringe for employees.
- High London salaries make the high cost of commuting affordable.
- The downside is crowded trains and congested roads.
- Settlements in the rural–urban fringe have become **dormitory towns and villages** – places where people sleep but are away during the day.

The rural-urban fringe

The rural-urban fringe is the change between rural and urban land uses. As you move out of urban London:

- the urban landscape gradually changes to a more rural one, but there are towns like Chelmsford along the way.
- bigger places like Chelmsford, gradually change to rural villages like Terling.

Six Second Summary

- Many villages close to London are seeing services decline.
- Villages in the rural–urban fringe are accessible, making it easy to get to London.
- Cities like London, and accessible rural areas, are interdependent.

Over to you

Summarise the economic, social and environmental costs and benefits of living in a village like Terling.

Student Book
See pages 192–193

You need to know:

- why some accessible rural areas experience economic and social change.

The biggest IT move in history!

In 2003 the Met Office moved from Bracknell, Berkshire to the edge of Exeter in East Devon. This resulted in:

- new jobs and a multiplier effect for this mainly rural region; a boost for the local economy.
- an extra £74 million a year to East Devon.

Why East Devon?

East Devon is attractive for companies as land costs are 10% of those in central London. Although 170 miles from London, it's very accessible.

- Daily flights from Exeter Airport to London and other cities.
- 0.5 km from the M5 junction 29.
- 42 daily train services to London.

This puts three pressures on East Devon.

1 Population change

- In 2014, Devon gained 5000 migrants from within the UK. 40% of them chose accessible East Devon (Figure 1).
- They included retired people, and families.

2 Pressure on housing

An increase in population increases demand for housing:

- Two-thirds of East Devon is an Area of Outstanding Natural Beauty (AONB). Planning permission for new housing is hard to get because of its impact on the scenery.
- Average incomes in East Devon are 10% below the UK average yet housing is only 3% cheaper than average.

East Devon – Inward and Outward migration, 2012–14

In or out migration	2012	2013	2014
Inward	7100	7000	7880
Outward	5600	5700	5800
Net	1500	1300	2080

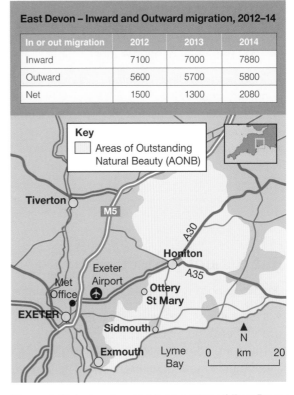

Figure 1 *Net migration and the location of East Devon*

3 Pressure on leisure and recreation

- Parts of the coastline are stunning.
- 30 minutes to the west is Dartmoor National Park.
- The accessibility of the area brings an estimated 15 million day trips a year!
- This puts pressure on roads and environmentally sensitive areas.

Draw a mind map to show **a)** who's moving to Devon, **b)** why they're moving to Devon, and **c)** the pressures they put on Devon.

- East Devon is an attractive area for relocation because land costs are cheaper and it has good transport links with London.
- East Devon is under pressure due to population change, demand for housing and the impact of visitors.

*Student Book
See pages 194–195*

You need to know:

- about the challenges that face changing rural areas.

Changing Cornwall

West of Devon lies Cornwall (Figure 1), home to about 540 000 people and, each August, four million tourists!

- Together with Devon, it's the UK's most popular holiday destination.
- It has a 700 km coastline with beaches, fishing harbours, and coves.
- It also has one of the UK's fastest growing populations.

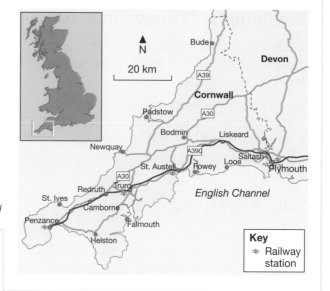

Figure 1 *Cornwall's location in south-west England*

But it's not perfect

Despite its holiday image, Cornwall has problems.

- It is nearly 140 km long. Rail transport is slow and there's no motorway.
- There are no population centres large enough to attract big employers. The largest, St Austell, has 23 000 people.
- There's no knowledge economy (except teleworkers) to raise incomes.
- The decline of Cornwall's traditional **primary** economy – farming, fishing, china clay quarrying and tin mining – has left few permanent, full-time jobs.
- Much employment is seasonal, part-time and low-paid.
- Cornwall has the UK's lowest weekly wage of £340, compared to a UK mean of £405.

Six Second Summary

- Despite its holiday image, Cornwall has economic problems.
- Primary employment has declined, incomes are low and it has few services.

Over to you

List five challenges facing Cornwall's economy.

Health and services

West Cornwall is one of the UK's most **deprived** areas (Figure 2). It has the UK's lowest average incomes, a high percentage of elderly people and few services.

- Only 38% of villages have a doctor's surgery; most only open one morning a week.
- Buses serve 70% of villages; there may be only 3–4 buses a day.
- The main hospital in Truro is, for many parts of west Cornwall, 50 km away.
- Some young people have to travel over 50 km for sixth form education or training.

Indicator of deprivation	What it measures	Proportion of population living in parts of Cornwall with this indicator
Income	People on low incomes	High
Employment	Those unable to work through unemployment, sickness or disability	Medium
Health and disability	People in poor physical or mental health	High
Education, skills and training	People with low educational attainment	Low
Housing and services	Affordability of housing and within reach of services e.g. transport, doctor	Very high
Crime	People affected by crime	Very low
Living environment	Those living in sub-standard housing (e.g. lack of heating, damp)	Very high

Figure 2 *The Index of Multiple Deprivation (IMD) in Cornwall*

Student Book
See pages 196–197

You need to know:

- about the benefits and costs of new opportunities in rural areas.

A day out in a quarry?

The Eden Project in Cornwall is a year-round, all-weather visitor attraction in what used to be a china clay quarry (Figure 1).

- It offers plant displays from around the world, plus winter ice skating, summer outdoor concerts, art projects etc.
- It was designed on sustainable principles, and an education centre runs courses on sustainability.
- It has generated £1 billion for the Cornish economy; employs 700 people directly, and created another 3000 jobs in associated companies (e.g. food suppliers).
- It is not as sustainable as hoped. It offers reduced admission for those arriving by public transport, but 97% use the car.
- Visitor numbers are falling.

Figure 1 *The Eden Project was visited by 13 million people in its first 10 years but there is evidence that, having been once, few return*

Farming and diversification

With farm incomes falling, farmers are trying to **diversify** (look for other activities to enable a farm to survive).

1 Farm shops

Until 2003, the Lobb brothers were earning just £30 000 from their farm.

- They developed a farm shop to sell their meat, vegetables and local produce (Figure 2).
- It was financed using EU and UK government grants.
- Their shop now has a turnover of over £700 000 per year and has created 20 jobs.
- There is a visitor centre with information on animal welfare and environmental farming, and it holds craft and food fairs.
- It's important regionally. Every £10 spent in farm shops is worth £23 to the local economy because of the multiplier effect.

Figure 2 *Lobb's Farm Shop, near the Lost Gardens of Heligan in Cornwall*

2 Tourist accommodation

- Using fields as camp sites or converting farm buildings into holiday cottages and leisure complexes.
- Barn conversions have led to a reduction in nesting places for birds.

Six Second Summary

- The Eden Project is an example of a new economic opportunity in a rural area.
- Farmers need to diversify to make an income.
- Farm shops and tourist accommodation are examples of diversification.
- New economic opportunities bring benefits, but may have environmental impacts.

Over to you

List the benefits to Cornwall of **a)** the Eden Project, **b)** farm shops, **c)** tourist accommodation.

Geographical skills

You need to know:

- how to use geographical skills to investigate south Cornwall
- how to use 4- and 6-figure grid references and interpret an OS map.

Identifying features of south Cornwall

The part of south Cornwall shown on Figure 1 contains

- the Eden Project (see Section 5.19)
- evidence of primary employment (see Section 5.18).

Big Idea

Maps can tell you about land use in an area.

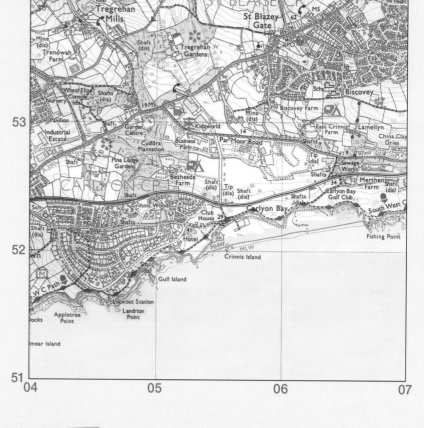

Figure 1 *1:25 000 Ordnance Survey (OS) map extract of part of south Cornwall*

Six Second Summary

- Text and symbols on a map provide clues about current and former land use.
- Contours provide information about height and relief.
- Spot heights give the height of a specific point.

Over to you

Look at Figure 1.

1 Give a 4-figure grid reference for the site of the Eden Project.
2 Give 4-figure grid references of other evidence that the area covered by the map is a tourist area.
3 Use 6-figure grid references to identify evidence of the following primary employment in this area: **a** mining, **b** farming.
4 Find Carlyon Bay. Planning permission was granted here in 2015 for the development of the derelict Cornwall Coliseum into 500 homes, restaurant and leisure complex. **a** Why do you think Carlyon Bay was chosen as a site for redevelopment? **b** Are there any disadvantages to the site?
5 What is the difference in height between the summit of the hill SSW of the Eden Project (marked by a spot height) and the Carlyon Bay Golf Club?

Topic 6
Geographical investigations

Your exam

- Topic 6 makes up Sections C1 and C2 in Paper 2.
- You choose one question within Sections C1 and C2, depending on which fieldwork topics you have done.
- In Section C1, choose **either** Investigating coastal change and conflict **or** Investigating river processes and pressures.
- In Section C2, choose **either** Investigating dynamic urban areas **or** Investigating changing rural areas.

> Tick these boxes to build a record of your revision

Your revision checklist

Fieldwork and research	Focus of your fieldwork	1	2	3
Investigating coastal change and conflict; Investigating river processes and pressures				
1 Formulating enquiry questions	• Understanding the kinds of questions capable of being investigated through fieldwork in **either** coastal **or** river environments			
2 Selecting fieldwork methods	Fieldwork data collection including: • one quantitative method to measure **either** how coastal management has affected beach morphology and sediment characteristics **or** changes in river channel characteristics			
	• one qualitative fieldwork method to collect data on **either** coastal management measures and their success **or** factors that might influence flood risk			
3 Secondary data sources	• **Either** for coasts, a geology map e.g. geology of Britain, **or** for rivers, a flood risk map such as that published by the Environment Agency			
	• For either, one other secondary source			
Investigating dynamic urban areas; Investigating changing rural areas				
1 Formulating enquiry questions	• Understanding the kinds of questions capable of being investigated through fieldwork in **either** urban **or** rural environments			
2 Selecting fieldwork methods	Fieldwork data collection including: • one quantitative method to collect data on perceptions of the quality of **either** urban **or** rural life			
	• one qualitative fieldwork method to collect data on environmental quality in **either** urban **or** rural areas			
3 Secondary data sources	• Census data for **either** urban **or** rural areas			
	• One other secondary source			

You need to know:

- how to investigate coastal processes and management using fieldwork and research.

Designing an enquiry

Coastal locations are good places for fieldwork enquiry as there are lots of questions that can be asked. Questions are usually the starting point for an enquiry. For Figure 1 you could ask:

- How did this beach get like this?
- Does this beach change in winter?
- What problems arise from the popularity of this beach?
- What impacts are all the people having here?

An enquiry is a series of stages, starting with a question (Figure 2, Stage 1) and ending with an answer or conclusion (Stage 5).

- You will probably have completed an enquiry in geography (or science) before and will have used **fieldwork** and practical work in the same way.
- Each stage is equally important, from the initial question, to the research and context, through to the overall conclusion and evaluation.
- At the end, you can reflect on what you have found and what it means.

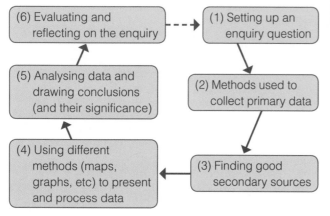

Figure 2 *The route to enquiry – a planning 'pathway' for your investigation*

Big Idea

Good fieldwork comes from designing and developing enquiry questions.

Enquiry and fieldwork

- Enquiry means the process of investigation to find an answer to a question – see Figure 2.
- Fieldwork means work carried out in the outdoors.

Figure 1 *This beach at Porthcurno, Cornwall, would be a good place for fieldwork!*

Six Second Summary

- A good enquiry depends on having a good question which is linked to coasts.
- Enquiry questions should not be too simple or too complex.
- To carry out a fieldwork enquiry about coasts, you need primary and secondary data.

Primary and secondary data

In order to answer your enquiry question, you will need to use both primary and secondary data.

- Fieldwork data which you collect yourself, or as part of a group. are called 'primary data'.
- Secondary data have been collected by someone else.

Over to you

Using Figure 1, write down and explain:

- one question that you could investigate using fieldwork along a stretch of coast
- two types of primary data that could help you answer this question
- one source of secondary data that could help you further.

Student Book
See pages 204–205

You need to know:

- how to use different techniques for collecting primary data in your coastal fieldwork.

Designing an enquiry

It's important to consider what data you need when you design your investigation, so that any data collected are as reliable and accurate as possible. Think about:

- **Sample size** – How many measurements will you be taking, and why? More measurements will generally mean more reliable conclusions.
- **Survey locations/sites** – Where will you collect the data, and how?
- **Accuracy** – How can you ensure that your data are accurate?

Figure 1 *Measuring pebble size along a beach, using callipers – an example of quantitative data*

Qualitative and quantitative data

There are two categories of data you might collect in a coastal environment:

- **Quantitative** data are data which can be measured (Figure 2), e.g. pebble size and length, or beach gradient.
- **Qualitative** data deals with opinions or judgments which are subjective. Examples include written accounts and photographs.

Data required	Equipment needed	Brief description and reasons for doing this
Beach gradient (in degrees)	Clinometer, tapes and ranging poles	The gradient is measured at distances up the beach from the water mark, often at 10 or 20 m intervals. The steepness of a beach can help us understand more about the processes operating and types of waves that often reach the coast.
Pebble shape (using a scale of roundness)	Identification chart	Measures how round or angular a sample of about 10–20 stones is. This tells us how eroded pebbles are – the more rounded they are, the more they have been smoothed off by abrasion.
Pebble size in cm or mm	Ruler or calliper (Figure 1)	Measures the length of the long axis of a sample of stones. This helps us know whether longshore drift is taking place, with the furthest material along a beach being the smallest.

Figure 2 *Examples of quantitative data used in beach investigations*

Types of sampling

When collecting quantitative data you need to decide on the type of **sampling** to use. Consider three main types in fieldwork for beach investigations:

- **Random** – where samples are chosen randomly, and every pebble has an equal chance of being selected.

- **Systematic** – using a regular system to collect data, for example every 10 m along a beach.
- **Stratified** – collecting a sample made up of different parts that reflect what the beach is like, e.g. selecting different sizes from a point on a beach where pebbles vary.

 Six Second Summary

- During a coastal fieldwork enquiry you will collect both qualitative and quantitative data.
- Quantitative data can be measured.
- Qualitative data are subjective.

 Over to you

Explain one way in which:

- you found your data might be unreliable
- you improved the reliability of your data collection.

Student Book
See pages 206–207

You need to know:

- how to present data using graphs, photos and maps.

Getting it together

- It's important to organise both your own data and group data so that it is ready for processing and presentation.
- Collect all your data (group and individual), collate it, select the data relevant to your study, then choose your presentation techniques.

Big Idea

Processing your data and choosing appropriate presentation techniques makes it clear and easy to interpret.

Presenting your data

Think more widely than just using bar charts, histograms and pie charts. Figure 1 shows a range of approaches to data presentation. You should think about:

- Why are you presenting the data in this way?
- Are the data **continuous**, e.g. sediment along a beach?

- Are the data in **categories?** Perhaps a bar chart can show these classifications (e.g. Figure 2 beach pebble data).
- Are your **sample sizes** different (e.g. 15 pebbles at one location, 17 at another)? You should turn raw numbers into percentages for each size category, then use a pie chart.

Maps / Cartography	GIS and photographs
• Used to show locations and patterns. • Mini-graphs and charts can be located on maps. • This makes it easier to compare patterns at locations.	• Used to show historic maps or sites which have been lost to erosion. • Useful for aerial photos of the coast to show land use. • Helps to show how places have changed after being affected by storms.
Table(s) of data	**Graphs and charts**
• Can be used to present raw data that you and your group collected. • Useful to highlight patterns and trends. • Can be highlighted and annotated, and can help to identify anomalies (any data which look unusual).	• There is a wide range of graphs and charts available. (Hint: make sure you choose the right chart, e.g. do you know when to use a pie chart or bar chart?) • Can show data and patterns clearly – easier to read than a table of data.

Figure 1 *Data presentation techniques need careful selection*

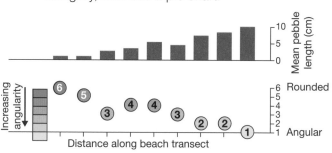

Figure 2 *These beach pebble data show both length and shape. They are category data*

Six Second Summary

- Organise your individual and group data so it is ready for processing.
- There are a wide range of techniques for presenting your data.
- Selecting the best technique makes your data clear and easy to interpret.

Other data presentation techniques

- Instead of just presenting graphs, locate them on a map or aerial photo (e.g. using Google Maps or ArcGIS Online). This makes differences easy to spot, and turns simple data into a geographical display.
- Annotated photographs and field sketches are good for showing evidence of processes, e.g. deposition, or property which is vulnerable to rapid coastal erosion.

Over to you

Using your own enquiry data:

- explain one technique you used to present beach sediment data
- sketch an example of one other presentation technique you used to show your results.

Student Book
See pages 208–209

You need to know:

- how to analyse your data and use the evidence in drawing your conclusions.

What is analysis?

To analyse your data you need to follow a sequence of:

- identifying patterns and trends in your results, and describing them
- making links between different sets of data
- identifying **anomalies** (unusual data which do not fit the general pattern of results)
- explaining reasons for patterns.

Cause and effect	Emphasis	Explaining	Suggesting
as a result of...	above all...	this shows...	could be caused by...
this results in...	mainly...	because...	this looks like...
triggering this...	mostly...	similarly...	points towards...
consequently...	most significantly...	therefore...	tentatively...
the effect of this is...	usually...	as a result of...	the evidence shows...

Figure 1 *The language of analysis*

When writing about your analysis, be clear and logical.
Figure 1 shows you examples of words and phrases that can help.

Quantitative analysis

Quantitative techniques are about handling numerical data from your enquiry. These can be analysed using statistical techniques, or, for example, a dispersion diagram such as Figure 2. Other techniques include calculating:

- the **mean** – the average value in a data set
- the **median** – the middle value in a data set when ranked in order
- the **mode** – the number that appears most frequently in the data set
- the **range** – the difference between the highest and lowest values
- **quartiles** – dividing a list of numbers into four equal groups – two above and two below the median. You could use quintiles (five groups) as well.

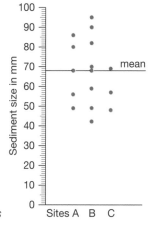

Figure 2 *A dispersion diagram showing the mean and variation in beach sediment sizes for three different sites*

Qualitative analysis

Photos and field sketches provide clues and evidence about fieldwork experience.

- Photos can be analysed using annotations.
- Field sketches can be used to analyse processes and change over time.

Writing a good conclusion

The conclusion is almost the end of your enquiry. In it you need to:

- refer back to the aim – what have you found out?
- quote the data that provide evidence
- comment on any unusual data that don't quite fit.

Six Second Summary

- Analysis shows patterns and trends in your data.
- There are a number of different quantitative analytical techniques you can use.
- Good analysis provides the evidence for strong conclusions.

Over to you

- Explain one way in which you analysed your beach fieldwork data.
- Explain one conclusion from your coastal enquiry.

Evaluating your coastal enquiry

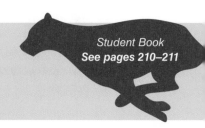

Student Book
See pages 210–211

You need to know:

- how to evaluate and think critically about different parts of your fieldwork enquiry.

Big Idea

Evaluation is the final – and perhaps most difficult – part of the enquiry process.

The importance of an evaluation

Evaluation can be tricky. You need to consider:

- how you collected your fieldwork data
- the reliability and accuracy of your results and conclusions.

It means more than simply a list of things that went wrong!

What might have affected your results?

There are two things to think about:

- **Reliability** – did your fieldwork investigation produce consistent results? In other words, if you were to repeat the enquiry, would you get the same results?
- **Validity** – did your enquiry produce conclusions that you trust?

Several factors influence the reliability and validity of your enquiry, as shown in Figure 1.

Possible sources of error	Impact on quality
Sample size	Smaller sample sizes usually means lower quality data.
Frequency of sample	Fewer sites reduces frequency, which then reduces quality.
Type of sampling	Sampling approaches may create 'gaps' and introduce bias in the results.
Equipment used	The wrong / inaccurate equipment can affect overall quality by producing incorrect results.
Time of survey	Different tides might influence beach accessibility and its measurable width.
Location of survey	Big variations in beach profiles and sediment characteristics can occur in locations close to each other.
Quality of secondary data	Age and reliability of secondary data affect their overall quality.

Figure 1 Sources of error in a coastal enquiry which can affect the reliability and validity of results

Being critical

Being critical means, simply, whether any shortcomings or limitations affected the results. Ask yourself these questions:

- How much do I trust the overall patterns or trends in my results?
- What is the chance that these patterns or trends could have happened by chance?
- Which of my conclusions is most reliable, compared to other conclusions?
- Which part of my enquiry created the most unreliable results, and why?

Link any knowledge gained from your enquiry to a theoretical model or idea. Think about key factors (Figure 2), then try to develop your own model.

Figure 2 Factors that may influence coastal processes

Six Second Summary

- Evaluation takes place at the end of an enquiry.
- It's important to identify the main sources of error.
- Be ready to accept that no study is ever perfect.

Over to you

- Explain one factor from your data collection that affected the reliability of your results.
- Explain one way that you could have improved the reliability of these results.

Student Book
See pages 212–213

You need to know:

• how to investigate river processes and management using fieldwork and research.

Designing an enquiry

River environments are good places for fieldwork enquiry as there are lots of questions that can be asked. Questions are usually the starting point for an enquiry. For Figure 1 you could ask:

• Why is this landscape like this?
• Does it ever change, e.g. between winter and summer?
• Why are the river valleys so deep?
• What impacts is farming having in the area?

An enquiry is a series of stages, starting with a question (Figure 2, Stage 1) and ending with an answer or conclusion (Stage 5).

• You will probably have completed an enquiry in geography (or science) before and will have used **fieldwork** and practical work in the same way.
• Each stage is equally important, from the initial question, to the research and context, through to the overall conclusion and evaluation.
• At the end, you can reflect on what you have found and what it means.

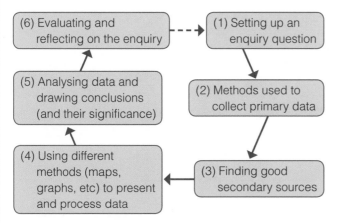

Figure 2 *The route to enquiry – a planning 'pathway' for your investigation*

Big Idea

Good fieldwork comes from designing and developing enquiry questions.

Enquiry and fieldwork

• Enquiry means the process of investigation to find an answer to a question – see Figure 2.
• Fieldwork means work carried out in the outdoors.

Figure 1 *This upland environment in the Peak District would be an interesting place for river fieldwork!*

Six Second Summary

• A good enquiry depends on having a good question which is linked to rivers.
• Enquiry questions should not be too simple or too complex.
• To carry out a fieldwork enquiry about rivers, you need primary and secondary data.

Primary and secondary data

In order to answer your enquiry question, you will need to use both primary and secondary data.

• Fieldwork data which you collect yourself, or as part of a group, are called 'primary data'.
• Secondary data have been collected by someone else.

Over to you

Using Figure 1, write down and explain:

• one question that you could investigate using fieldwork along a stretch of river
• two types of primary data that could help you answer this question
• one source of secondary data that could help you further.

Student Book
See pages 214–215

You need to know:

• how to use different techniques for collecting primary data in your river fieldwork.

Designing an enquiry

It's important to consider what data you need when you design your investigation, so that any data collected are as reliable and accurate as possible. Think about:

• **Sample size** – How many measurements will you be taking, and why? More measurements will generally mean more reliable conclusions.
• **Survey locations/sites** – Where will you collect the data, and how?
• **Accuracy** – How can you ensure that your data are accurate?

Figure 1 *Measuring pebble size from a river, using a homemade pebblometer – an example of quantitative data*

Qualitative and quantitative data

There are two categories of data you might collect in a river environment:

• **Quantitative** data are data which can be measured (Figure 2), e.g. pebble size and length, or stream width.
• **Qualitative** data deal with opinions or judgments which are subjective. Examples include written accounts and photographs.

Data required	Equipment needed	Brief description and reasons for doing this
River gradient (in degrees)	Clinometer, tapes and ranging poles	The gradient is measured at sites along the river and is measured over 10 or 20m. The gradient of a river can, for example, help us understand more about the processes operating and the influence of geology.
River speed (velocity)	Flow meter or float	Measures how fast the river is flowing. This tells us about the amount of energy in a river and we can investigate whether it conforms to a model.
Pebble size in cm or mm	Ruler or pebbleometer/ calliper (Figure 1)	Measures the length of the long axis of a sample of stones. This helps to link processes of river erosion with position within a river catchment.
River width and depth	Tape measure, ruler	Measures width of the river in cm. Measures depth in cm by taking five readings – one at each bank, plus at a quarter, half, and three-quarters of the width.

Figure 2 *Examples of quantitative data used in river investigations*

Types of sampling

When collecting quantitative data you to need decide on the type of **sampling** to use. Consider three main types in fieldwork for river investigations:

• **Random** – where samples are chosen randomly, and every pebble has an equal chance of being selected.

• **Systematic** – using a regular system to collect data, for example every 20 cm across a river.
• **Stratified** – collecting a sample made up of different parts; e.g., selecting different pebble sizes from a point in the river to include the range of pebble sizes found there.

 Six Second Summary

• During a river fieldwork enquiry you will collect both qualitative and quantitative data.
• Quantitative data can be measured.
• Qualitative data are subjective.

 Over to you

Explain one way in which:

• you found your data might be unreliable
• you improved the reliability of your data collection.

Student Book
See pages 216–217

You need to know:

- how to present data using graphs, photos and maps.

Getting it together

- It's important to organise both your own data and group data so that it is ready for processing and presentation.
- Collect all your data (group and individual), collate it, select the data relevant to your study, then choose your presentation techniques.

Big Idea

Processing your data and choosing appropriate presentation techniques makes it clear and easy to interpret.

Presenting your data

Think more widely than just using bar charts, histograms and pie charts. Figure 1 shows a range of approaches to data presentation. You should think about:

- Why are you presenting the data in this way?
- Are the data **continuous**, e.g. sediment along a river?

- Are the data in **categories?** Perhaps a bar chart can show these classifications (e.g. Figure 2 river pebble data).
- Are your **sample sizes** different (e.g. 15 pebbles at one location, 17 at another)? You should turn raw numbers into percentages for each size category, then use a pie chart.

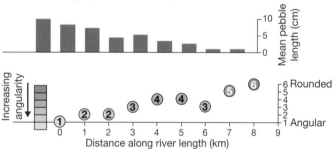

Figure 2 *These river pebble data show both length and shape. They are category data.*

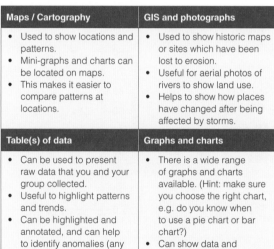

Maps / Cartography	GIS and photographs
• Used to show locations and patterns. • Mini-graphs and charts can be located on maps. • This makes it easier to compare patterns at locations.	• Used to show historic maps or sites which have been lost to erosion. • Useful for aerial photos of rivers to show land use. • Helps to show how places have changed after being affected by storms.
Table(s) of data	**Graphs and charts**
• Can be used to present raw data that you and your group collected. • Useful to highlight patterns and trends. • Can be highlighted and annotated, and can help to identify anomalies (any data which look unusual).	• There is a wide range of graphs and charts available. (Hint: make sure you choose the right chart, e.g. do you know when to use a pie chart or bar chart?) • Can show data and patterns clearly – easier to read than a table of data.

Figure 1 *Data presentation techniques need careful selection*

Six Second Summary

- Organise your individual and group data so it is ready for processing.
- There are a wide range of techniques for presenting your data.
- Selecting the best technique makes your data clear and easy to interpret.

Other data presentation techniques

- Instead of just presenting graphs, locate them on a map or aerial photo (e.g. using Google Maps or ArcGIS Online). This makes differences easy to spot, and turns simple data into a geographical display.
- Annotated photographs and field sketches are good for showing evidence of processes, e.g. deposition, or an area of a valley where it is liable to flood.

Over to you

Using your own enquiry data:

- explain one technique you used to present river sediment data
- sketch an example of one other presentation technique you used to show your results.

Student Book
See pages 218–219

You need to know:

- how to analyse your data and use the evidence in drawing your conclusions.

What is analysis?

To analyse your data you need to follow a sequence of:

- identifying patterns and trends in your results, and describing them
- making links between different sets of data
- identifying **anomalies** (unusual data which do not fit the general pattern of results)
- explaining reasons for patterns.

When writing about your analysis, be clear and logical. Figure 1 shows you examples of words and phrases that can help.

Cause and effect	Emphasis	Explaining	Suggesting
as a result of...	above all...	this shows...	could be caused by...
this results in...	mainly...	because...	this looks like...
triggering this...	mostly...	similarly...	points towards...
consequently...	most significantly...	therefore...	tentatively...
the effect of this is...	usually...	as a result of...	the evidence shows...

Figure 1 *The language of analysis*

Quantitative analysis

Quantitative techniques are about handling numerical data from your enquiry. These can be analysed using statistical techniques, or, for example, a dispersion diagram such as Figure 2. Other techniques include calculating:

- the **mean** – the average value in a data set
- the **median** – the middle value in a data set when ranked in order
- the **mode** – the number that appears most frequently in the data set
- the **range** – the difference between the highest and lowest values
- **quartiles** – dividing a list of numbers into four equal groups – two above and two below the median. You could use quintiles (five groups) as well.

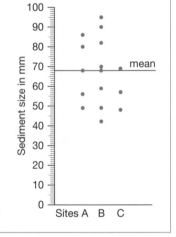

Figure 2 *A dispersion diagram showing the mean and variation in river sediment sizes for three different sites*

Qualitative analysis

Photos and field sketches provide clues and evidence about fieldwork experience.

- Photos can be analysed using annotations.
- Field sketches can be used to analyse processes and change over time.

Writing a good conclusion

The conclusion is almost the end of your enquiry. In it you need to:

- refer back to the aim - what have you found out?
- quote the data that provide evidence
- comment on any unusual data that don't quite fit.

Six Second Summary

- Analysis shows patterns and trends in your data.
- There are a number of different quantitative analytical techniques you can use.
- Good analysis provides the evidence for strong conclusions.

Over to you

- Explain one way in which you analysed your river fieldwork data.
- Explain one conclusion from your river enquiry.

Student Book
See pages 220–221

- how to evaluate and think critically about different parts of your fieldwork enquiry.

Big Idea

Evaluation is the final – and perhaps most difficult – part of the enquiry process.

The importance of an evaluation

Evaluation can be tricky. You need to consider:

- how you collected your fieldwork data
- the reliability and accuracy of your results and conclusions.

It means more than simply a list of things that went wrong!

What might have affected your results?

There are two things to think about:

- **Reliability** – did your fieldwork investigation produce consistent results? In other words, if you were to repeat the enquiry, would you get the same results?
- **Validity** – did your enquiry produce conclusions that you trust?

Several factors influence the reliability and validity of your enquiry, as shown in Figure 1.

Possible sources of error	Impact on quality
Sample size	Smaller sample sizes usually means lower quality data.
Frequency of sample	Fewer sites reduces frequency, which then reduces quality.
Type of sampling	Sampling approaches may create 'gaps' and introduce bias in the results.
Equipment used	The wrong / inaccurate equipment can affect overall quality by producing incorrect results.
Time of survey	Different times of the year will significantly influence the amount of water in the river and may not be representative.
Location of survey	Big variations in river channel depth and width, as well as sediment characteristics can occur in locations close to each other.
Quality of secondary data	Age and reliability of secondary data affect their overall quality.

Figure 1 *Sources of error in a river enquiry which can affect the reliability and validity of results*

Being critical

Being critical means, simply, whether any shortcomings or limitations affected the results. Ask yourself these questions:

- How much do I trust the overall patterns or trends in my results?
- What is the chance that these patterns or trends could have happened by chance?
- Which of my conclusions is most reliable, compared to other conclusions?
- Which part of my enquiry created the most unreliable results, and why?

Link any knowledge gained from your enquiry to a theoretical model or idea. Think about key factors (Figure 2), then try to develop your own model.

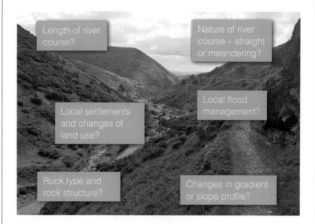

Figure 2 *Factors that may influence river processes*

Six Second Summary

- Evaluation takes place at the end of an enquiry.
- It's important to identify the main sources of error.
- Be ready to accept that no study is ever perfect.

Over to you

- Explain one factor from your data collection that affected the reliability of your results.
- Explain one way that you could have improved the reliability of these results.

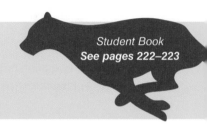

Student Book
See pages 222–223

You need to know:

- how to investigate variations in urban quality of life using fieldwork and research.

Designing an enquiry

Urban environments are good places for fieldwork enquiry as there are lots of questions that can be asked. Questions are usually the starting point for an enquiry. For Figure 1 you could ask:

- How might the numbers of people change over the course of a day?
- Why are there buildings of different ages?
- Have people tried to improve the quality of this area?
- How might this area change in the future?

An enquiry is a series of stages, starting with a question (Figure 2, Stage 1) and ending with an answer or conclusion (Stage 5).

- You will probably have completed an enquiry in geography (or science) before and will have used **fieldwork** and practical work in the same way.
- Each stage is equally important, from the initial question, to the research and context, through to the overall conclusion and evaluation.
- At the end, you can reflect on what you have found and what it means.

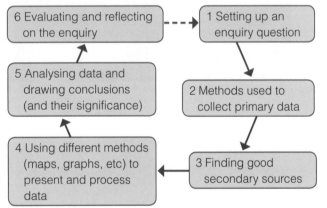

Figure 2 *The route to enquiry – a planning 'pathway' for your investigation*

Big Idea

Good fieldwork comes from designing and developing enquiry questions.

Enquiry and fieldwork

- Enquiry means the process of investigation to find an answer to a question – see Figure 2.
- Fieldwork means work carried out in the outdoors.

Figure 1 *This urban environment in central Sheffield has lots of opportunities for fieldwork*

Six Second Summary

- A good enquiry depends on having a good question which is linked to urban places.
- Enquiry questions should not be too simple or too complex.
- To carry out a fieldwork enquiry about urban areas, you need primary and secondary data.

Primary and secondary data

In order to answer your enquiry question, you will need to use both primary and secondary data.

- Fieldwork data which you collect yourself, or as part of a group, are called 'primary data'.
- Secondary data have been collected by someone else.

Over to you

Using Figure 1, write down and explain:

- one question that you could investigate using fieldwork in an urban environment
- two types of primary data that could help you answer this question
- one source of secondary data that could help you further.

Student Book
See pages 224–225

You need to know:

- how to use different techniques for collecting primary data in your urban fieldwork.

Designing an enquiry

It's important to consider what data you need when you design your investigation, so that any data collected are as reliable and accurate as possible. Think about:

- **Sample size** – How many measurements will you be taking, and why? More measurements will generally mean more reliable conclusions.
- **Survey locations/sites** – Where will you collect the data, and how?
- **Accuracy** – How can you ensure that your data are accurate?

Figure 1 *Using a camera to record the built environment at the coast – an example of qualitative data*

Qualitative and quantitative data

There are two categories of data you might collect in an urban environment:

- **Quantitative** data are data which can be measured (Figure 2), e.g. pedestrian counts and traffic surveys.
- **Qualitative** data deal with opinions or judgments which are subjective. Examples include written accounts and photographs.

Data required	Equipment needed	Brief description and reasons for doing this
Land use map	Large-scale base map and land use map key	Categories of land use are recorded either along a line (e.g. sides of a road) or in areas to produce a spatial picture of urban land use.
Shopping / environmental quality survey	Pre-prepared environmental quality survey	Measures different characteristics of a place based on numerical judgements, with a simple scoring system that can be tallied at the end.
'Local or visitor' coded questionnaire	Questionnaire with 'closed' questions and specific answers	Specific questions are used to gauge the perceptions of people (respondents), either visitors or locals, about how they 'feel' about an area.

Figure 2 *Examples of quantitative data used in urban investigations*

Types of sampling

When collecting quantitative data you need to decide on the type of **sampling** to use. Consider three main types in fieldwork for urban investigations:

- **Random** – where samples are chosen randomly, and every person has an equal chance of being selected.

- **Systematic** – using a regular system to collect data, for example every 50 m along a street.
- **Stratified** – collecting a sample made up of different parts; e.g., selecting different ages of people to include the range of people in the town.

Six Second Summary

- During an urban fieldwork enquiry you will collect both qualitative and quantitative data.
- Quantitative data can be measured.
- Qualitative data are subjective.

Over to you

Explain one way in which:

- you found your data might be unreliable
- you improved the reliability of your data collection.

Student Book
See pages 226–227

You need to know:

- how to present data using graphs, photos and maps.

Getting it together

- It's important to organise both your own data and group data so that it is ready for processing and presentation.
- Collect all your data (group and individual), collate it, select the data relevant to your study, then choose your presentation techniques.

Big Idea

Processing your data and choosing appropriate presentation techniques makes it clear and easy to interpret.

Presenting your data

Think more widely than just using bar charts and histograms. Figure 1 shows approaches to data presentation, and Figure 2 a land use map. You should think about:

- Why are you presenting the data in this way?
- Are the data **continuous**, e.g. numbers of people / pedestrians along a transect?

- Are the data in **categories?** If so, a bar chart can show these classifications.
- Are your **sample sizes** different (e.g. an environmental quality score of 15/20 at one location, 17/25 at another)? If so, turn raw numbers into percentages, then use a pie chart.

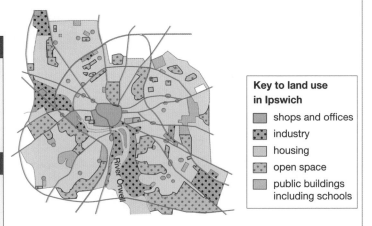

Key to land use in Ipswich

- ▨ shops and offices
- ▨ industry
- ☐ housing
- ▨ open space
- ▨ public buildings including schools

Figure 2 *Land use maps are a common data presentation method*

Maps / Cartography	GIS and photographs
• Used to show locations and patterns. • Mini-graphs and charts can be located on maps. • This makes it easier to compare patterns at locations.	• Used to show historic maps to show change in an urban area. • Useful for aerial photos of the town / city to show land use. • Helps to show deprivation and / or 'health' of a place.
Table(s) of data	**Graphs and charts**
• Can be used to present raw data that you and your group collected. • Useful to highlight patterns and trends. • Can be highlighted and annotated, and can help to identify anomalies (any data which look unusual).	• There is a wide range of graphs and charts available. (Hint: make sure you choose the right chart, e.g. do you know when to use a pie chart or bar chart?) • Can show data and patterns clearly – easier to read than a table of data.

Figure 1 *Data presentation techniques need careful selection*

Six Second Summary

- Organise your individual and group data so it is ready for processing.
- There are a wide range of techniques for presenting your data.
- Selecting the best technique makes your data clear and easy to interpret.

Other data presentation techniques

- Instead of just presenting graphs, locate them on a map or aerial photo (e.g. using Google Maps or ArcGIS Online). This makes differences easy to spot, and turns simple data into a geographical display.
- Annotated photographs and field sketches are good for showing evidence of processes, e.g. deprivation.

Over to you

Using your own enquiry data:

- Explain one technique you used to present environmental quality data.
- Sketch an example of one other presentation technique you used to show your results.

*Student Book
See pages 228–229*

You need to know:

- how to analyse your data and use the evidence in drawing your conclusions.

What is analysis?

To analyse your data you need to follow a sequence of:

- identifying patterns and trends in your results, and describing them
- making links between different sets of data
- identifying **anomalies** (unusual data which do not fit the general pattern of results)
- explaining reasons for patterns.

When writing about your analysis, be clear and logical.
Figure 1 shows you examples of words and phrases that can help.

Cause and effect	Emphasis	Explaining	Suggesting
as a result of…	above all…	this shows…	could be caused by…
this results in…	mainly…	because…	this looks like…
triggering this…	mostly…	similarly…	points towards…
consequently…	most significantly…	therefore…	tentatively…
the effect of this is…	usually…	as a result of…	the evidence shows…

Figure 1 *The language of analysis*

Quantitative analysis

Quantitative techniques are about handling numerical data from your enquiry. These can be analysed using statistical techniques, or, for example, a dispersion diagram such as Figure 2. Other techniques include calculating:

- the **mean** – the average value in a data set
- the **median** – the middle value in a data set when ranked in order
- the **mode** – the number that appears most frequently in the data set
- the **range** – the difference between the highest and lowest values
- **quartiles** – dividing a list of numbers into four equal groups – two above and two below the median. You could use quintiles (five groups) as well.

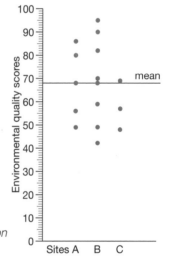

Figure 2 *A dispersion diagram showing the mean and variation in environmental quality scores for three different urban sites*

Qualitative analysis

Photos and field sketches provide clues and evidence about fieldwork experience.

- Photos can be analysed using annotations.
- Field sketches can be used to analyse processes and change over time.

Writing a good conclusion

The conclusion is almost the end of your enquiry. In it you need to:

- refer back to the aim – what have you found out?
- quote the data that provide the evidence
- comment on any unusual data that don't quite fit.

Six Second Summary

- Analysis shows patterns and trends in your data.
- There are a number of different quantitative analytical techniques you can use.
- Good analysis provides the evidence for strong conclusions.

Over to you

- Explain one way in which you analysed your urban fieldwork data.
- Explain one conclusion from your urban enquiry.

Evaluating your urban enquiry

Student Book
See pages 230–231

You need to know:

- how to evaluate and think critically about different parts of your fieldwork enquiry.

Big Idea

Evaluation is the final – and perhaps most difficult – part of the enquiry process.

The importance of an evaluation

Evaluation can be tricky. You need to consider:

- how you collected your fieldwork data
- the reliability and accuracy of your results and conclusions.

It means more than simply a list of things that went wrong!

What might have affected your results?

There are two things to think about:

- **Reliability** – did your fieldwork investigation produce consistent results? In other words, if you were to repeat the enquiry, would you get the same results?
- **Validity** – did your enquiry produce conclusions that you trust?

Several factors influence the reliability and validity of your enquiry, as shown in Figure 1.

Possible sources of error	Impact on quality
Sample size	Smaller sample sizes usually means lower quality data.
Frequency of sample	Fewer sites reduces frequency, which then reduces quality.
Type of sampling	Sampling approaches may create 'gaps' and introduce bias in the results.
Equipment used	The wrong / inaccurate equipment can affect overall quality by producing incorrect results.
Time of survey	Different days or times of day might influence perceptions and pedestrian flow, for example.
Location of survey	Big variations in environmental quality can occur between places very close to each other.
Quality of secondary data	Age and reliability of secondary data affect their overall quality.

Figure 1 *Sources of error in an urban enquiry which can affect the reliability and validity of results*

Being critical

Being critical means, simply, whether any shortcomings or limitations affected the quality of the results. Ask yourself:

- How much do I trust the overall patterns or trends in my results?
- What is the chance that these patterns or trends could have happened by chance?
- Which of my conclusions is most reliable, compared to other conclusions?
- Which part of my enquiry created the most unreliable results, and why?

Link any knowledge gained from your enquiry to a theoretical model or idea. Think about key factors (Figure 2), then try to develop your own model.

Figure 2 *Factors that may influence urban deprivation and quality of life*

Six Second Summary

- Evaluation takes place at the end of an enquiry.
- It's important to identify the main sources of error.
- Be ready to accept that no study is ever perfect.

Over to you

- Explain one factor from your data collection that affected the reliability of your results.
- Explain one way that you could have improved the reliability of these results.

Student Book
See pages 232–233

Designing an enquiry

Rural environments are good places for fieldwork enquiry as there are lots of questions that can be asked. Questions are usually the starting point for an enquiry. For Figure 1 you could ask:

- What might the challenges be for people living in these areas?
- Why is there such a range of different landscapes?
- Have people tried to improve the quality of this area?
- How might this place change in the future?

An enquiry is a series of stages, starting with a question (Figure 2, Stage 1) and ending with an answer or conclusion (Stage 5).

- You will probably have completed an enquiry in geography (or science) before and will have used **fieldwork** and practical work in the same way.
- Each stage is equally important, from the initial question, to the research and context, through to the overall conclusion and evaluation.
- At the end, you can reflect on what you have found and what it means.

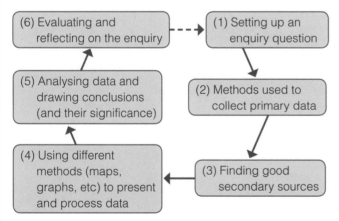

Figure 2 *The route to enquiry – a planning 'pathway' for your investigation*

(6) Evaluating and reflecting on the enquiry → (1) Setting up an enquiry question → (2) Methods used to collect primary data → (3) Finding good secondary sources → (4) Using different methods (maps, graphs, etc) to present and process data → (5) Analysing data and drawing conclusions (and their significance) → (6) Evaluating and reflecting on the enquiry

Big Idea

Good fieldwork comes from designing and developing enquiry questions.

Enquiry and fieldwork

- Enquiry means the process of investigation to find an answer to a question – see Figure 2.
- Fieldwork means work carried out in the outdoors.

Figure 1 *This rural environment near Keswick, in the Lake District, has lots of opportunities for fieldwork*

Six Second Summary

- A good enquiry depends on having a good question which is linked to rural places.
- Enquiry questions should not be too simple or too complex.
- To carry out a fieldwork enquiry about rural places, you need primary and secondary data.

Primary and secondary data

In order to answer your enquiry question, you will need to use both primary and secondary data.

- Fieldwork data which you collect yourself, or as part of a group, are called 'primary data'.
- Secondary data have been collected by someone else.

Over to you

Using Figure 1, write down and explain

- one question that you could investigate using fieldwork in a rural area
- two types of primary data that could help you answer this question
- one source of secondary data that could help you further.

Primary data collection for rural fieldwork

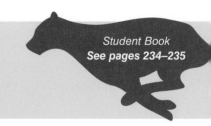

*Student Book
See pages 234–235*

- how to use different techniques for collecting primary data in your rural fieldwork.

Designing an enquiry

It's important to consider what data you need when you design your investigation, so that any data collected are as reliable and accurate as possible. Think about:

- **Sample size** – How many measurements will you be taking, and why? More measurements will generally mean more reliable conclusions.
- **Survey locations/sites** – Where will you collect the data, and how?
- **Accuracy** – How can you ensure that your data are accurate?

Figure 1 Fast traffic can be a hazard and lower the quality of life for people living in a village – an example of qualitative data

Qualitative and quantitative data

There are two categories of data you might collect in a rural environment:

- **Quantitative** data are data which can be measured (Figure 2), e.g. pedestrian counts and traffic surveys.
- **Qualitative** data deal with opinions or judgments which are subjective. Examples include written accounts and photographs.

Data required	Equipment needed	Brief description and reasons for doing this
Land use map	Large-scale base map and land use map key	Categories of land use are recorded either along a line (e.g. sides of a road) or in areas to produce a spatial picture of rural land use.
Environmental quality survey	Pre-prepared environmental quality survey	Measures different characteristics of a place based on numerical judgements, with a simple scoring system that can be tallied at the end.
'Local or visitor' coded questionnaire	Questionnaire with 'closed' questions and specific answers	Specific questions are used to gauge the perceptions of people (respondents), either visitors or locals, about how they 'feel' about an area.

Figure 2 Examples of quantitative data used in rural investigations

Types of sampling

When collecting quantitative data you need to decide on the type of **sampling** to use. Consider three main types in fieldwork for rural investigations:

- **Random** – where samples are chosen randomly, and every person has an equal chance of being selected.

- **Systematic** – using a regular system to collect data, e.g. every 20 m along a village street.
- **Stratified** – collecting a sample made up of different parts; e.g., selecting different ages of people to include the range of people in a village.

- During a rural fieldwork enquiry you will collect both qualitative and quantitative data.
- Quantitative data can be measured.
- Qualitative data are subjective.

Over to you

Explain one way in which:

- you found your data might be unreliable
- you improved the reliability of your data collection.

Student Book
See pages 236–237

- how to present data using graphs, photos and maps.

Getting it together

- It's important to organise both your own data and group data so that it is ready for processing and presentation.
- Collect all your data (group and individual), collate it, select the data relevant to your study, then choose your presentation techniques.

 **Big Idea**

Processing your data and choosing appropriate presentation techniques makes it clear and easy to interpret.

Presenting your data

Think more widely than just using bar charts, histograms and pie charts. Figure 1 shows a range of approaches to data presentation. You should think about:

- Why are you presenting the data in this way?
- Are the data **continuous**, e.g. numbers of people / pedestrians along a transect?

- If the data are in **categories** then a bar chart can show these classifications.
- Are your **sample sizes** different (e.g. an environmental quality score of 15/20 at one location, 17/25 at another)? If so, turn raw numbers into percentages, then use a compound bar chart (Figure 2).

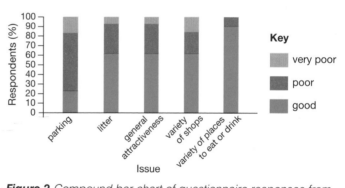

Figure 2 *Compound bar chart of questionnaire responses from a village survey*

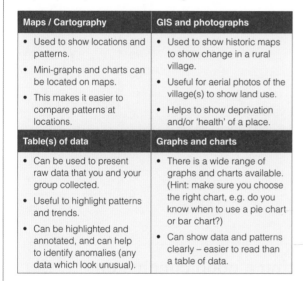

Maps / Cartography	GIS and photographs
• Used to show locations and patterns. • Mini-graphs and charts can be located on maps. • This makes it easier to compare patterns at locations.	• Used to show historic maps to show change in a rural village. • Useful for aerial photos of the village(s) to show land use. • Helps to show deprivation and/or 'health' of a place.
Table(s) of data	**Graphs and charts**
• Can be used to present raw data that you and your group collected. • Useful to highlight patterns and trends. • Can be highlighted and annotated, and can help to identify anomalies (any data which look unusual).	• There is a wide range of graphs and charts available. (Hint: make sure you choose the right chart, e.g. do you know when to use a pie chart or bar chart?) • Can show data and patterns clearly – easier to read than a table of data.

Figure 1 *Data presentation techniques need careful selection*

Other data presentation techniques

- Instead of just presenting graphs, locate them on a map or aerial photo (e.g. using Google Maps or ArcGIS Online). This makes differences easy to spot, and turns simple data into a geographical display.
- Annotated photographs and field sketches are good for showing evidence of processes, e.g. deprivation, or an area of a village that has been enhanced.

 **Six Second Summary**

- Organise your individual and group data so it is ready for processing.
- There are a wide range of techniques for presenting your data.
- Selecting the best technique makes your data clear and easy to interpret.

 **Over to you**

Using your own enquiry data:

- explain one technique you used to present environmental quality data
- sketch an example of one other presentation technique you used to show your results.

Analysis and conclusions – rural enquiry

You need to know:

- how to analyse your data and use the evidence in drawing your conclusions.

Student Book
See pages 238–239

What is analysis?

To analyse your data you need to follow a sequence of:

- identifying patterns and trends in your results, and describing them
- making links between different sets of data
- identifying **anomalies** (unusual data which do not fit the general pattern.
- explaining reasons for patterns.

Cause and effect	Emphasis	Explaining	Suggesting
as a result of…	above all…	this shows…	could be caused by…
this results in…	mainly…	because…	this looks like…
triggering this…	mostly…	similarly…	points towards…
consequently…	most significantly…	therefore…	tentatively…
the effect of this is…	usually…	as a result of…	the evidence shows…

Figure 1 *The language of analysis*

When writing about your analysis, be clear and logical.
Figure 1 shows you examples of words and phrases that can help.

Quantitative analysis

Quantitative techniques are about handling numerical data from your enquiry. These can be analysed using statistical techniques, or, for example, a dispersion diagram such as Figure 2. Other techniques include calculating:

- the **mean** – the average value in a data set
- the **median** – the middle value in a data set when ranked in order
- the **mode** – the number that appears most frequently in the data set
- the **range** – the difference between the highest and lowest values
- **quartiles** – dividing a list of numbers into four equal groups – two above and two below the median. You could use quintiles (five groups) as well.

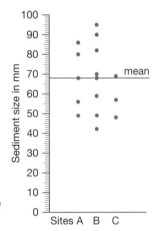

Figure 2 *A dispersion diagram showing the mean and variation in environmental quality scores for three different villages*

Qualitative analysis

Photos and field sketches provide clues and evidence about fieldwork experience.

- Photos can be analysed using annotations.
- Field sketches can be used to analyse processes and change over time.

Writing a good conclusion

The conclusion is almost the end of your enquiry. In it you need to:

- refer back to the aim – what have you found out?
- quote the data that provide the evidence
- comment on any unusual data that don't quite fit.

Six Second Summary

- Analysis shows patterns and trends in your data.
- There are a number of different quantitative analytical techniques you can use.
- Good analysis provides the evidence for strong conclusions.

Over to you

- Explain one way in which you analysed your rural fieldwork data.
- Explain one conclusion from your rural enquiry.

Evaluating your rural enquiry

Student Book
See pages 240 –241

- how to evaluate and think critically about different parts of your fieldwork enquiry.

 Big Idea

Evaluation is the final – and perhaps most difficult – part of the enquiry process.

The importance of an evaluation

Evaluation can be tricky. You need to consider:

- how you collected your fieldwork data
- the reliability and accuracy of your results and conclusions.

It means more than simply a list of things that went wrong!

What might have affected your results?

There are two things to think about:

- **Reliability** – did your fieldwork investigation produce consistent results? In other words, if you were to repeat the enquiry, would you get the same results?
- **Validity** – did your enquiry produce conclusions that you trust?

Several factors influence the reliability and validity of your enquiry, as shown in Figure 1.

Possible sources of error	Impact on quality
Sample size	Smaller sample sizes usually means lower quality data.
Frequency of sample	Fewer sites reduces frequency, which then reduces quality.
Type of sampling	Sampling approaches may create 'gaps' and introduce bias in the results.
Equipment used	The wrong / inaccurate equipment can affect overall quality by producing incorrect results.
Time of survey	Different days or time of day might influence perceptions and tourist numbers, for example.
Location of survey	Big variations in environmental quality can occur between places very close to each other.
Quality of secondary data	Age and reliability of secondary data affect their overall quality.

Figure 1 *Sources of error in a rural enquiry which can affect the reliability and validity of results*

Being critical

Being critical means, simply, whether any shortcomings or limitations affected the quality of the results. Ask yourself:

- How much do I trust the overall patterns or trends in my results?
- What is the chance that these patterns or trends could have happened by chance?
- Which of my conclusions is most reliable, compared to other conclusions?
- Which part of my enquiry created the most unreliable results, and why?

Link any knowledge gained from your enquiry to a theoretical model or idea. Think about key factors (Figure 2), then try to develop your own model.

Figure 2 *Factors that may influence rural deprivation and quality of life*

Six Second Summary

- Evaluation takes place at the end of an enquiry.
- It's important to identify the main sources of error.
- Be ready to accept that no study is ever perfect.

Over to you

- Explain one factor from your data collection that affected the reliability of your results.
- Explain one way that you could have improved the reliability of these results.

Topic 7
People and the biosphere

Your exam

- Topic 7 People and the biosphere is part of Paper 3, People and environment issues – making geographical decisions.

- Paper 3 is a 90-minute written exam and makes up 25% of your final grade. The whole paper carries 64 marks (including 4 marks for SPaG).

- You must answer all parts of Paper 3. Section A contains questions on Topic 7 People and the biosphere, Section B on Topic 8 Forests under threat (pages 138–149), and Section C on Topic 9 Consuming energy resources (pages 150–163). Section D is a decision-making question which draws together all three topics.

Tick these boxes to build a record of your revision

Your revision checklist

Spec Key Idea	Detailed content that you should know	1	2	3
7.1 The Earth is home to a number of very large ecosystems (biomes), the distribution of which is affected by climate and other factors	• How the global distribution and characteristics of major biomes are influenced by climate			
	• Local factors can alter biome distribution locally, and how biotic and abiotic components of biomes interact			
7.2 The biosphere is a vital life-support system for people as it provides both goods and services	• How the biosphere provides resources for indigenous and local people but is also increasingly exploited commercially for energy, water and mineral resources			
	• How the biosphere regulates the composition of the atmosphere, maintains soil health and regulates water within the hydrological cycle, providing globally important services			
	• Global and regional trends – increasing demand for food, energy and water resources and theories on relationships between population and resources			

- how the global distribution of biomes is affected by climate.

*Student Book
See pages 244–245*

Getting to know the biosphere

- It is the living layer of the Earth's surface between the rocks (lithosphere) and the air (atmosphere).
- It contains all plants and animals.
- It is divided into nine large regions called **biomes** (large-scale ecosystems).

Biomes

Each biome has its own climate, plants and animals. Figure 1 shows their distribution. The location and characteristics of each are determined by climate because it affects the growth of plants:

- temperature determines the length of growing season
- precipitation provides water
- sunshine hours affect photosynthesis.

Temperature

Temperature and sunshine intensity are controlled by **latitude**.

- Near the Equator, the sun is more intense and at a high angle all year.
- Towards the Poles, winter is longer and colder; the climate is more seasonal.
- In polar areas, sunshine intensity is low; plant growth is limited.

Precipitation

Precipitation is also influenced by latitude.

- There are three convection cells north and south of the Equator (Section 1.3).
- In the rising parts of these cells, precipitation is high, as air pressure is low.
- In the descending parts of these cells, precipitation is low, as air pressure is high.

Key	Vegetation
Tundra	Grasses, lichens and dwarf shrubs; no trees
Coniferous forest	Coniferous trees, e.g. pine
Temperate deciduous forest	Deciduous trees, e.g. oak
Temperate grassland	Short or tall grasses and few trees
Mediterranean	Evergreen and deciduous trees and shrubs
Desert	Cacti and succulents, but few of them
Tropical rainforest	Evergreen trees growing all year round
Tropical grassland (savanna)	Grass with some trees, e.g. acacia
Other biomes (e.g. polar, ice, mountains)	

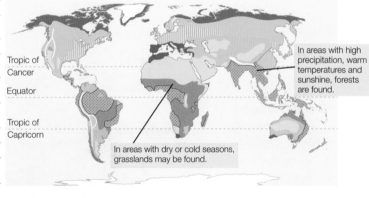

In areas with high precipitation, warm temperatures and sunshine, forests are found.

In areas with dry or cold seasons, grasslands may be found.

Figure 1 *Biomes are arranged in 'belts' around the Earth at different **latitudes** (how far north or south a location is from the Equator)*

 Six Second Summary

- The biosphere is divided into nine major biomes.
- The location and characteristics of each biome is determined by climate.

 Over to you

Explain in one sentence, why each of the following biomes is found in the location shown in Figure 1:
a) tundra **b)** hot desert **c)** temperate grassland.

Student Book
See pages 246–247

You need to know:

- how local factors affect biome location and type
- how biotic and abiotic parts of biomes interact.

UK ecosystems

If the UK had no urban areas or farming it would a) be covered by temperate forest biome (deciduous trees, e.g. oak), b) produce different **ecosystems** (localised biomes) affected by **local factors**. These are explained below.

How local factors affect ecosystems

1 Rock and soil type

- The acidity/alkalinity of soil influences the types of plants that will grow.
- Rocks release nutrients and chemicals into the soil through chemical weathering.
- Soils can depend on rock type – they can be acidic, neutral, or alkaline.

2 Water availability and drainage

Some plants prefer wet soil, others dry soil. How wet the soil is depends on:

- the amount of precipitation
- the amount of evaporation
- how permeable the soil is.

3 Altitude

- Temperatures drop as height increases.
- Freezing temperatures are common at high altitudes.
- Rainfall increases with height.

These factors form a pattern called **altitudinal zonation**.

Biotic or abiotic?

Biomes consist of:

- the **biotic** (living) part – plant and animal life
- the **abiotic** (non-living) part – the atmosphere, water, rock and soil.

These parts are linked together (Figure 1). Local changes in abiotic parts can change the conditions for biotic parts.

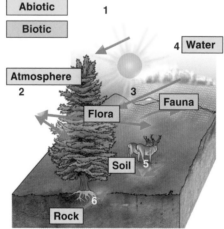

1 Energy is provided by photosynthesis.

2 Plants take in CO_2 and release oxygen, animals do the opposite. Nitrogen is also exchanged.

3 Energy flows through the food web from plants to herbivores and carnivores, then to detritivores and decomposers that consume dead plants and animals.

4 Water moves through the soil, plants and animals and back into the atmosphere via respiration and evaporation.

5 When plants and animals die, decomposition returns nutrients to the soil.

6 Weathering of rock provides soil nutrients, e.g. phosphates.

Figure 1 *The biotic and abiotic parts of ecosystems are linked together*

 Six Second Summary

- Local factors (rock and soil type, water availability and altitude) affect the plant and animal species in a biome.
- Local changes in abiotic parts of ecosystems can affect biotic parts.

Over to you

Draw a spider diagram to show how local factors can affect the plant and animal species in a biome.

Geographical skills: learning about climate and biomes

Student Book
See pages 248–249

You need to know:

- how to analyse climate data and interpret climate graphs.

1 Understanding temperate forest and tundra biomes

Nottinghamshire has a climate typical of the **temperate deciduous forest** biome. The climate data for Nottingham (latitude 53°N) in Figure 1 show:

- no months with temperatures below freezing
- year-round rainfall
- year-round sunshine hours.

By contrast, Barrow, Alaska (located at latitude 71°N) shown in Figure 2, has a very different climate:

- temperatures do not go above 5°C
- it is too cold for trees to grow
- the biome is **tundra** (cold desert)
- the sun disappears for 2 months between November and January
- there is 24-hour daylight between May and July.

Months	J	F	M	A	M	J	J	A	S	O	N	D
Precipitation (mm)	61	47	50	54	52	63	58	62	59	71	66	69
Temperature (°C)	4.0	4.1	6.3	8.4	11.6	14.5	16.7	16.5	14.0	10.4	6.7	4.2
Sunshine (hours)	55	73	104	141	187	171	191	180	131	99	64	49

Figure 1 *Climate averages for Nottingham, UK*

Months	J	F	M	A	M	J	J	A	S	O	N	D
Temperature (°C)	-25.2	-25.7	-24.7	-16.8	-6.1	2.0	4.9	3.9	0.1	-8.2	-17.4	-22.1
Precipitation (mm)	3	3	2	4	5	8	25	27	18	10	5	3

Figure 2 *Climate averages for Barrow, Alaska. There are no recorded sunshine data for Barrow.*

2 Understanding tropical grassland biomes

Tropical grassland climate is different to both temperate deciduous forest and tundra. Figure 3 shows the climate graph for a tropical grassland area in Kenya, Africa and the vegetation that can be found there.

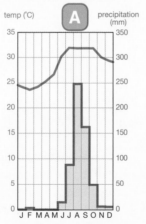

Figure 3 *Climate graph for Kenya (**A**) and a photo of vegetation found in the tropical grassland biome (**B**)*

 Six Second Summary

- Climate data and climate graphs can be used to understand contrasting biomes.

 Over to you

1   Analyse the differences in the climate of the temperate deciduous forest and tundra by calculating: **a)** the total annual precipitation, **b)** the monthly temperature range.

2   Write three statements about the climate of the tropical grassland biome using Figure 3.

You need to know:

Student Book
See pages 250–251

- how biomes provide goods and services for people
- how resources are being exploited commercially from biomes.

Valuing biomes

Biomes are estimated to be worth US$4–6 trillion each year, as they provide humans with different types of **goods** and **services** (Figure 1).

- Goods are physical materials.
- Services are functions.
- The importance of **ecosystem services** varies – they can be globally important (maintaining a healthy atmosphere) or locally important (depended upon by local people).

Provisioning services (goods)	Supporting services
These are products obtained from ecosystems: • food: nuts, berries, fish, game, crops • fuelwood • timber for buildings and other uses • genetic and chemical material.	These keep the ecosystem healthy so it can provide the other services: • nutrient cycling • photosynthesis and food webs • soil formation.
Regulating services	**Cultural services**
These services link to other physical systems and keep areas, and the whole planet, healthy: • storing carbon, and emitting oxygen, which keep the atmosphere in balance • purifying water and regulating the flow of water within the hydrological cycle.	These are benefits people get from visiting, or living in, a healthy ecosystem: • recreation and tourism • education and science • spiritual well-being and happiness.

Figure 1 Ecosystem services can be divided into four categories

Destroying ecosystem services

Commercial exploitation of biomes provides profit for TNCs, jobs and income. Large areas of biomes are cleared for:

- commercial farming, including cattle and crops
- mining metal ores
- timber
- construction of dams for HEP and water supply.

Once cleared for commercial exploitation, the biome cannot grow back.

- It cannot store carbon, prevent flooding or provide recreation.
- Many ecosystem services are destroyed.

Figure 2 shows the percentage of four biomes destroyed by human activity.

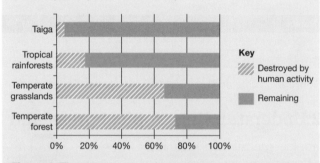

Figure 2 The destroyed and remaining percentage of four biomes

Ecosystem services for indigenous peoples

Around 30 000 Efe people of the Ituri tropical rainforest in the Congo Basin of Africa depend on the rainforest for:

- wood and leaves to build small, temporary houses
- wood for cooking fires
- monkey, antelope, fish, yams, nuts, mushrooms and berries for eating
- giant forest hog for selling and trading
- plants and wild honey for traditional medicine
- their faith – they worship the rainforest itself.

Six Second Summary

- Biomes provide goods and services.
- Local indigenous peoples depend directly on ecosystem services.
- Ecosystem services are being exploited commercially.

 **Over to you**

List: **a)** the benefits of commercial exploitation of biomes, **b)** the drawbacks of commercial exploitation of biomes.

Student Book
See pages 252–253

You need to know:

- how biomes maintain the atmosphere, soil health and the hydrological cycle.

Brilliant biomes!

Without biomes, and especially forests:

- climate would be very different
- flooding would be more common
- the level of carbon dioxide in the atmosphere would rise
- soil would be unhealthy.

Healthy air

Biomes are an important **carbon sink**.

- They remove carbon dioxide from the atmosphere.
- They store carbon by locking it up in biotic material (about 120 billion tonnes each year). This is called **carbon sequestration** and works through **photosynthesis**.
- Biomes store carbon as biomass e.g. leaves, branches, animal tissue.
- When plants and animals die, the dead biomass ends up in the soil – making soil a carbon sink.
- If humans destroy biomes and burn biomass, carbon is released into the atmosphere.
- An increase in carbon dioxide in the atmosphere has been linked to global warming.

Healthy soils

Biomes are important in maintaining soil health.

- Soil health (or fertility) is maintained by the **nutrient cycle** (Figure 1).
- Nutrient cycles can be easily disrupted.
- Removing biomass (e.g. logging timber) takes away a large nutrient store.
- Heavy rain can wash away litter.
- Deforested areas are at risk from soil erosion.

Water supply and flood risk

Biomes are an important part of the **hydrological cycle**, and destroying a forest biome can have serious impacts:

- interception is reduced and soil is eroded
- there is much less infiltration into the soil making groundwater supplies low
- surface run-off increases and flooding becomes more frequent
- soil dries out quickly, so overall evaporation is reduced, creating a drier climate.

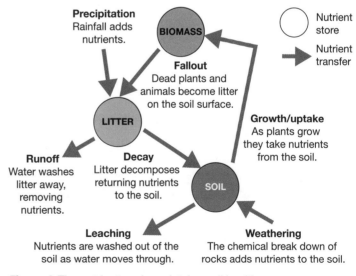

Figure 1 *The nutrient cycle maintains soil health*

 Six Second Summary

- Biomes maintain healthy air, soils and the hydrological cycle.

Over to you

Explain how biomes maintain healthy air, in less than 50 words.

Student Book
See pages 254–255

You need to know:

- how population growth and industrialisation have increased the demand for resources.

More people, more demand

Rising demand for **natural resources** (food, energy, water) has negative effects on the biosphere.

- Biomes are destroyed for land for farming, housing, factories and mining.
- Key species are removed as a result of hunting and obtaining timber.
- Rivers and the atmosphere are polluted.

Pressure on natural resources has grown between 1975 and 2015, due to:

- rising global population (4.1 to 7.3 billion) – more people need more food and water.
- rising affluence (average US$ 3700 to US$10 400 per person) – wealthier people use more energy.
- increasing urbanisation (38% to 55%) – towns and cities have sprawled over biomes increasing the demand for space, food and water.

Industrialisation

Since the 1970s, many countries have been through **industrialisation**, and people have moved from working on farms in the countryside, to working in factories and offices in cities.

- This has been most noticeable in Asia, e.g. China and India.
- Construction and resource consumption has risen.
- Transport, e.g. high speed trains and car use, have grown rapidly in China and India.
- Thailand's urban population doubled between 1990 and 2015.

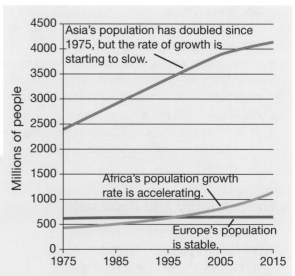

Figure 1 *Population growth in Asia, Africa and Europe*

Asian growth

The population of Asia has grown by 78% since 1975. Figure 1 shows how population growth has varied around the world.

- Resource consumption in Asia has increased rapidly (Figure 2).
- It's a challenge for the Earth to provide enough resources for more people.

People are on average richer, so use far more energy.

Asia	Water consumption (cubic km)	Oil consumption (millions of tonnes)	Coal consumption (millions of tonnes)	Meat consumption (million tonnes)
1975	1000	453	388	22
2015	1700	1450	2810	135
Percentage increase 1975 – 2015	70%	220%	624%	513%

As people get richer, they eat more meat.

Figure 2 *The increased demand for resources in Asia*

Population growth, industrialisation, urbanisation and rising wealth has led to a dramatic increase in demand for resources.

Six Second Summary

- Population growth, industrialisation, urbanisation and rising wealth has led to a dramatic increase in demand for resources, particularly in Asia.

Over to you

Create a mind map to show why population growth and industrialisation have increased demand for resources.

You need to know:

*Student Book
See pages 256–257*

- the theories on the relationship between population and resources.

2030: Perfect storm?

The '2030 perfect storm' idea is that by 2030 the world will be running out of resources. Geographers have worried about this for 200 years. There are two viewpoints – those of Malthus and those of Boserup (Figure 1).

View	Idea	People who held this view	Theory
Pessimistic	Population will grow so large that the planet will run out of resources, leading to a crisis.	Malthusians or Neo-Malthusians: - Thomas Malthus - Professor John Beddington - The Club of Rome	In 1798, Reverend Thomas Malthus argued that: • population would increase geometrically (1,2,4,8,16 etc) • food production could only increase arithmetically (1,2,3,4,5, etc) • population would outstrip food supply leading to a 'population versus resource crisis' • population would have to fall by '**positive** checks' (war, famine) or '**preventative** checks' (later marriage, fewer children). The Club of Rome made this view popular again in the 1970s as: • there have been local food crises • population growth is projected to reach 12 billion.
Optimistic	As population grows, humans invent new technologies to allow more resources to be supplied.	Boserupians: - Ester Boserup	In 1965, Danish economist, Ester Boserup argued: • 'necessity is the mother of invention' • as population grows, humans will invent new ways to produce food e.g. technology such as farm machinery, GM crops and irrigation. More recently: • food production has become more technically advanced • renewable and sustainable resources might support people.

Figure 1 *The theories of Malthus and Boserup compared*

Future population growth

It is important, but difficult, to predict future population growth.

- Since 2000, estimates have increased.
- It was thought global population would peak at 9 billion in 2050.
- Now it looks likely to grow beyond 2050.
- There is a large margin of uncertainty.

Six Second Summary

- There are optimistic and pessimistic views on the relationship between population and resources.
- Future population growth is hard to predict.
- Quality, rather than quantity of resources may create a 'crisis'.

Quality or quantity?

A 'crisis' might occur because the **quality** of resources becomes too low (rather than **quantity** becoming too small).

- Rivers and lakes could be too polluted.
- Soils may be eroded so few crops can grow.
- We may turn to burning very dirty, polluting coal.
- The price of resources may become too high for some.
- The very poor and hungry may never access the resources they need.

Over to you

Summarise the views of Malthus and Boserup, using one sentence for each about: **a)** their belief, **b)** what they thought happened to resources, **c)** evidence to show they're right or wrong.

Topic 8
Forests under threat

Your exam

- Topic 8 Forests under threat is part of Paper 3, People and environment issues – making geographical decisions.

- Paper 3 is a 90-minute written exam and makes up 25% of your final grade. The whole paper carries 64 marks (including 4 marks for SPaG).

- You must answer all parts of Paper 3. Section A contains questions on Topic 7 People and the biosphere (pages 130–137), Section B on Topic 8 Forests under threat, and Section C on Topic 9 Consuming energy resources (pages 150–163). Section D is a decision-making question which draws together all three topics.

Tick these boxes to build a record of your revision

Your revision checklist

Spec Key Idea	Detailed content that you should know	1	2	3
8.1 The structure, functioning and adaptations of the tropical rainforest reflect the equatorial climate	• How biotic and abiotic characteristics are interdependent, and how plants and animals are adapted to the climate			
	• Why tropical rainforests have a very high rate of nutrient cycling which, in turn, supports high levels of biodiversity and complex food webs			
8.2 The taiga shows different characteristics, reflecting the more extreme and highly seasonal climate	• How biotic and abiotic characteristics are interdependent, how taiga plants and animals are adapted to the climate			
	• Why the taiga has lower productivity, with less active nutrient cycling and much lower levels of biodiversity			
8.3 Tropical rainforests are threatened directly by deforestation and indirectly by climate change	• Causes of deforestation: logging; subsistence and commercial agriculture; demand for fuel wood, biofuels and mineral resources; HEP			
	• Why climate change is an indirect threat to the health of tropical rainforests			
8.4 The taiga is increasingly threatened by commercial development	• Direct threats (logging, pulp and paper production) and indirect threats (mineral exploitation, fossil fuels and HEP potential)			
	• How acid precipitation, forest fires, pests and diseases contribute to a loss of biodiversity			
8.5 Conservation and sustainable management of tropical rain forests are vital if goods and services are not to be lost for future generations	• Advantages and disadvantages of global actions (CITES, REDD) designed to protect rainforest species and areas, and why deforestation rates are rising in some areas but falling in others			
	• The challenge of sustainable forest management; why alternative livelihoods could better protect remaining rainforests			
8.6 The taiga wilderness areas need to be protected from over-exploitation	• Challenges of creating and maintaining protected wilderness, national parks and sustainable forestry in the taiga			
	• Reasons for conflicting views on protecting or exploiting forest and natural resources in the taiga			

You need to know:

- how plants and animals are adapted to the climate in the tropical rainforest.

Student Book
See pages 260–261

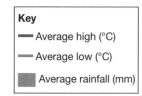

Biodiversity booms!

Biodiversity in tropical rainforests is high; 10% of all the world's plant and animal species live in the Amazon rainforest. Biodiversity is so high in tropical rainforests because:

- of the climate. They receive heat, water and sunlight all year
- species have evolved over thousands of years
- multiple layers of vegetation provide lots of different habitats.

Forest structure

All rainforests have a typical structure with layers of plants and trees (Figure 2).

Adapting to the climate

Plant Adaptations	Animal Adaptations
Evergreen hardwood trees (e.g. Mahogany) • Tall, slender trunks. • Leaves and branches only at top. • Buttress roots support weight of trees.	**Birds** (e.g. parrots and macaws) • Loud calls to easily attract mates. • Powerful beaks to break open nuts.
Lianas • Climbing plants that use trees as their 'trunk'. • Their stems cling to trees and climb up to the sunlight in the canopy.	**Primates** (e.g. lemurs and monkeys) • Live in the canopy where most food is. • Long tails used for balance. • Strong claws grip trees and branches.

Figure 3 *Examples of how animals and plants have adapted to rainforests*

Rainforests: climate

Tropical rainforests are found where there is an **equatorial climate** (e.g. areas of South America, Figure 1). This means that:

- rain falls every month – at least 60 mm
- temperatures are high all year around – between 26–32°C.

Figure 1 *Climate graph for Belem in Brazil. The average temperature is a near-constant 32°C with over 2900 mm of rainfall a year.*

Key
— Average high (°C)
— Average low (°C)
▮ Average rainfall (mm)

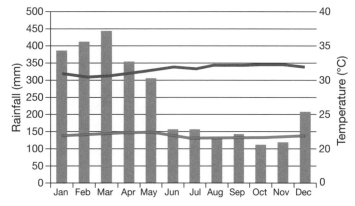

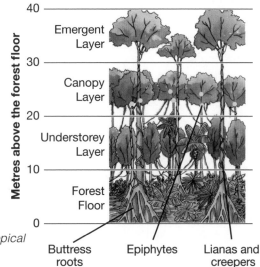

Hardwood, evergreen trees break through the dense canopy layer to reach the sunlight. Monkeys and birds live here.

Home to tree snakes, birds, tree frogs and other animals, lots of food is available.

Contains young trees and those with large leaves to capture sunlight; huge numbers of insects live here.

Dark forest floor-shade-loving ferns with large leaves live here plus mammals like jaguars.

Figure 2 *The layered structure of the tropical rainforest*

Six Second Summary

- Biodiversity in rainforests is high.
- The rainforest has a layered structure.
- Plants and animals have adapted to the rainforest environment.

Over to you

Make a large simple copy of Figure 2. Annotate each layer with the different plants and animals (and their adaptations) that would be found there.

You need to know:

- about the relationships between nutrient cycles, biodiversity and food webs in tropical rainforests.

Student Book
See pages 262–263

Nutrient cycles

Nutrient cycles show how nutrients (chemicals needed by plants to grow), are moved around biomes. Nutrients circulate between abiotic (non-living) and biotic (living) parts of biomes.

The nutrient cycle in rainforests

In tropical rainforests nutrient recycling is rapid. The cycle differs from that shown on page 135. It has:

- A **larger biomass store** because of the dense vegetation.
- A **smaller litter store** because **decay** happens quickly in the hot, wet conditions, returning nutrients to the soil.
- A **larger take-up of nutrients** from the soil because plants grow quickly all year round.
- A **larger supply of nutrients** from **chemical weathering** which is faster in hot wet conditions.
- A **larger loss of nutrients** due to the constant flow of moisture through the soil.

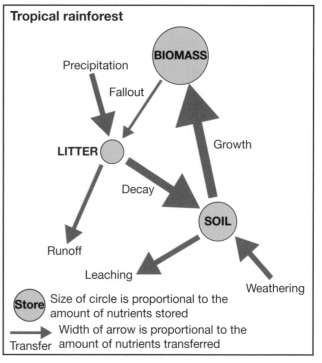

Tropical rainforest

Store	Size of circle is proportional to the amount of nutrients stored
Transfer	Width of arrow is proportional to the amount of nutrients transferred

Figure 1 *The nutrient cycle in a tropical rainforest. Remember that this process is rapid – large volumes of nutrients move quickly between stores.*

The impact of deforestation

- Destroys biomass – where most of the nutrients are stored.
- Litter and soil are easily eroded by heavy rains – no longer protected by the forest.
- Land cannot be farmed for long as rainforest soil contains so few nutrients.

The web of life

Plants and animals in biomes are connected through **food webs**.

- Sunlight provides the basic energy source.
- Plants convert sunlight into carbohydrates through photosynthesis.
- As one organism feeds on another, energy passes between them.

Food webs in rainforests

Complex food webs are found in tropical rainforests as animals are selective about what they eat. Food webs represent a delicate balance between species, but this balance can be easily disrupted and reduce biodiversity.

Six Second Summary

- Nutrient cycling is rapid in tropical rainforests.
- This supports high levels of biodiversity and complex food webs.
- The nutrient cycle can be easily disrupted e.g. by deforestation.

Over to you

Define the keywords mentioned above: biomass, nutrients, decay and chemical weathering.

Student Book
See pages 264–265

You need to know:

- how plants and animals are adapted to life in the taiga
- why productivity is lower in the taiga than in tropical rainforests.

Taiga: climate

- Short, wet summers that last for 3 months.
- Long, cold winters.
- Low rainfall (350–750 mm per year).
- Snow lies on the ground for many months.

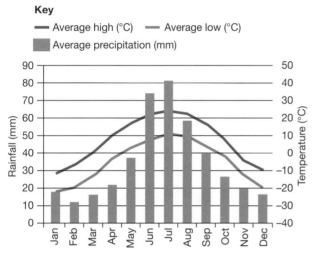

Figure 1 *Climate graph for Fort McMurray in Alberta, Canada – an area of taiga biome*

Adapting to extremes

Biodiversity is low. Plants and animals can only survive with special adaptations to cope with the cold.

- Mammals have **thick, oily fur** to retain heat.
- Some animals **hibernate** in winter.
- Some birds and animals **migrate**, only arriving in the taiga in the warmer months.
- Trees are **coniferous** (evergreen). There is only really one layer of vegetation.

Six Second Summary

- The taiga climate is harsh and biodiversity is low.
- Plants and animals have adapted to the climate.
- The nutrient cycle is slower, with smaller stores and flows than in the rainforest.

Nutrient cycles in the taiga

The nutrient cycle occurs more slowly than in rainforests.

- The stores are smaller, with smaller flows of nutrients between them.

In addition:

- precipitation is lower
- chemical weathering is limited by the cold
- most nutrients are in the litter because decay happens slowly in cold temperatures
- the biomass store is small because trees only grow for a few months of the year.

Taiga nutrient cycle

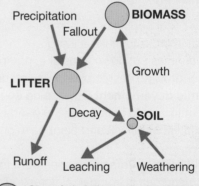

 Store Size of circle is proportional to the amount of nutrients stored

Transfer Width of arrow is proportional to the amount of nutrients transferred

Figure 2 *The nutrient cycle in the taiga*

Productivity in the taiga

Net Primary Productivity (NPP) measures how much biomass (plant and animal growth) is added to a biome each year.

- It is measured in grams per square metre.
- NPP is higher in rainforests with more sunlight, higher temperatures and rainfall.
- NPP is lower in the taiga where it is colder with a dry season.

Over to you

Picture, or research, a pine tree. How do their leaves and pine cones help them survive in the taiga climate?

*Student Book
See pages 266–267*

You need to know:

- how tropical rainforests are being destroyed by human activity.

Threats to rainforest biomes

People threaten forest biomes directly and indirectly.

- **Direct threats** involve deliberate cutting down of trees (**deforestation**) for timber, to build roads or convert forest to farmland.
- **Indirect threats** come from pollution, global warming or disease (see Section 8.5).

Direct threats

Deforestation is the main direct threat. There are a number of causes (Figure 1). The points below are reasons for deforestation.

- **Poverty** Local people cut down trees to farm the land, in order to make a living.
- **Debt** Countries cut down forests, exporting timber or growing cash crops to pay off debts.
- **Economic development** Developing countries cut down forests to build roads, expand cities and build HEP stations and plant palm oil for biodiesel.
- **Demand for resources** Countries want to exploit resources in tropical rainforests. These resources include timber, oil, gas, iron ore and gold.

Using GIS to identify rainforest loss

Satellite technology and GIS (e.g. Google Earth) is making it much easier to find out how much tropical rainforest is being cut down.

- Satellite images show that deforestation first happens along major roads
- It continues along smaller tracks leading from the roads until whole areas are cleared.

Rates of deforestation

Different countries have different rates and causes of deforestation.

Country	Annual Rate of deforestation (%)	Causes of different rates
Burundi	–4.7	**Poverty** - cutting down trees provides income.
Indonesia	–1.9	The **palm oil industry**.
Malaysia	–0.7	Increasing environmental **protection** is lowering deforestation rates.
Brazil	–0.6	
Democratic Republic of Congo	–0.2	**Isolation** means that the forest is hard to access and cut down, so deforestation rate Is low.

Figure 2 The annual rates of deforestation and reasons for this, in five countries

Logging, legal and illegal 3%

Fires, mining, urbanisation, road construction, dams 3%

Small-scale, subsistence agriculture 33%

Large-scale commercial agriculture including soybeans 1%

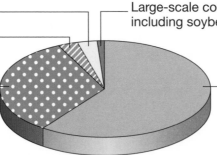

Cattle ranches 60%

Figure 1 The direct causes of deforestation in the Amazon rainforest. Notice that cattle ranching is the main cause.

 Six Second Summary

- Deforestation is the biggest direct threat to tropical rainforests.
- There are many causes of deforestation including cattle ranching, agriculture and demand for resources.
- Rates of deforestation differ between countries.

Over to you

Create a spider diagram showing threats to tropical rainforests.

Student Book
See pages 268–269

You need to know:

- how tropical rainforests are being threatened by climate change.

Global warming: an indirect threat

Climate change is the greatest indirect threat to rainforests.

- Population increase and rising demand for resources adds greenhouse gases to the atmosphere (see Section 1.8)
- There is already evidence of the impacts of climate change. Figure 1 summarises expected changes as temperatures rise.

Temperature rise	Impact on species and biomes
1°C	10% of land species face extinction. Alpine, mountain and tundra biomes shrink as temperatures rise.
2°C	15–40% of land species face extinction. Biomes begin to shift towards the Poles; animal migration patterns and breeding times change. Extreme weather at unusual times of the year affect pollination and migration.
3°C	20–50% of land species face extinction. Forest biomes are stressed by drought; fire risk increases on grassland. Flooding causes loss of mangroves. Pests and diseases thrive, e.g. bark beetles, which destroy coniferous forests.

Figure 1 *The impact of rising temperatures on species and biomes. Global warming is occurring too rapidly for many species to adapt; they may become extinct.*

It's not all bad!

The rate of deforestation in the Amazon has slowed since 2004 because:

- an area the size of France has been protected
- demand for resources fell from 2008 because of the global recession (though it rose again after 2014)
- the government has cracked down on illegal logging
- more Brazilians have voted for the Green Party.

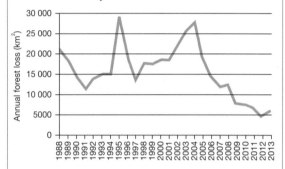

Figure 2 *Annual forest loss in Amazonia 1988–2013. Notice the reduction since 2004*

Climate stress

Evidence for a changing climate can be seen in the Amazon rainforest.

- Three severe droughts in 2005, 2010 and 2014–15 were unusually close together.
- During these droughts, the Amazon started emitting CO_2 instead of absorbing it. Plant growth slowed, and forest fires broke out.
- The forest ecosystem was put under stress as leaf litter dried, causing decomposer organisms to die.
- Leaves in the canopy died, reducing food supply and affecting food webs.

Some scientists believe that:

- deforestation is making droughts more common and severe
- fewer trees means less evaporation and transpiration, so fewer clouds and less rain
- if drought becomes more common, rainforests will die back
- dieback could accelerate global warming as rainforests become sources of CO_2 rather than carbon sinks.

Six Second Summary

- Climate change is the main indirect threat to tropical rainforests.
- During droughts the Amazon rainforest can switch from absorbing CO_2 to emitting it.
- The rate of deforestation in the Amazon rainforest has slowed since 2004.

Over to you

Create a table with two columns for points that **a)** agree and **b)** disagree with the statement: 'Global warming is probably the biggest single threat to tropical rainforests'.

Big Idea

Commercial exploitation involves deforestation to access resources.

Deforestation

Global Forest Watch reported in 2013 that countries with the greatest deforestation are those with taiga forests.

• Canada and Russia accounted for over 40% of all deforestation between 2000 and 2013 (Figure 1).

But less is heard about taiga deforestation than rainforest destruction; the reasons for this are below.

• The biome is vast. Despite deforestation, only 8% of the taiga has been lost.
• The taiga is in remote regions, so it's 'out of sight'.
• There are few 'cute and cuddly' endangered animals.

Country	Type of forest	Percentage (%) of global forest deforestation 2000 – 2013
Canada	Taiga (boreal forest)	21.4
Russia	Taiga (boreal forest)	20.4
Brazil	Tropical rainforest	14.2
USA	Taiga (boreal forest)	6.1
Bolivia	Tropical rainforest	4.2
Indonesia	Tropical rainforest	3.7

Figure 1 *Deforestation of all intact forests between 2000 and 2013. Most deforestation takes place in taiga forests, rather than tropical rainforests.*

Softwood, pulp and paper

Paper comes from softwood trees (e.g. fir and pine). The world uses about 400 million tonnes of paper a year, causing deforestation.

• 80% of all trees cut down each year are softwoods.
• Softwood is also used for construction timber (e.g. for roofs, window and door frames) and board.

Minerals, fossil fuels and HEP

• Clearing forest allows access to minerals (e.g. diamonds, gold, iron ore) and fossil fuels (oil and gas).
• There are over 4000 mines in Ontario, Canada alone. Each one means taiga is destroyed.
• Building dams for HEP generation also destroys taiga.

Figure 2 shows the impact of deforestation on an area cleared for tar sands mining.

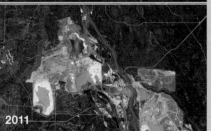

Figure 2 *The impact of tar sands mining near Fort McMurray, Canada, 1984–2011. Notice the tracks and roads; tailing ponds (light grey areas containing toxic waste); reduction in size of the forest.*

 Six Second Summary

• Deforestation is highest in countries with taiga forests.
• Causes of deforestation include the use of timber, and accessing mineral resources and fossil fuels (including tar sands).

Over to you

Draw a spider diagram to show how different activities threaten the taiga.

- how the taiga is indirectly threatened by forest fires, pests and diseases, and acid rain.

Forest fire

How can forest fires be a problem in the cold and wet taiga?

- Summers can be hot and dry. During storms, lightning can start fires.
- Pine needles form a thick layer - perfect tinder for a fire.
- The resin inside coniferous trees can burn easily.

Forest fires have been increasing since the 1990s. Whilst some fires are good (allowing the taiga ecosystem to regenerate), too many can cause long-term reductions in biodiversity.

- Trees cannot mature between fires, and so the forest does not regenerate properly.
- Species that *can* withstand fire begin to dominate, reducing biodiversity.
- Trees that cannot withstand fire decline, as do bird and insect populations that feed on them.

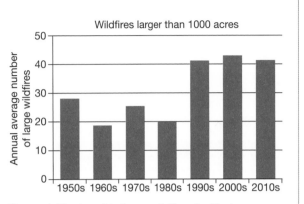

Figure 1 The trend in large wildfires in Alaska

Figure 1 shows the trend in **wildfires** in Alaska which have increased since the 1990s. This may be linked to global warming and hotter, drier summers, with an increase in fires as temperatures rise.

Pests and diseases

Normally, very cold winters kill off disease-carrying insects and their larvae. However, with warmer temperatures, there has been an increase in insect infestation and disease in coniferous trees.

Pests and diseases have several consequences.

- The wood cannot be sold as timber, reducing the commercial value of forests.
- The food web changes as some tree species are killed, changing the ecosystem.
- Landscapes change from dense forest to more open mixed grass and woodland, with fewer trees.
- Biodiversity is reduced.

Acid rain

Acid rain (rain with a pH lower than 5.7) forms when:

- burning fossil fuels release sulphur dioxide and nitrogen dioxide into the air
- these gases react with water in the clouds
- precipitation falls as low concentrations of sulphuric and nitric acids to the ground.

Acid rain gradually weakens trees by reducing:

- their ability to photosynthesise due to damaged needles
- essential alkali plant nutrients (e.g. calcium) in the soil.

Weaker trees become vulnerable to disease and pests, and biodiversity is reduced as the taiga become stressed.

 Six Second Summary

- The taiga is under threat from the increase in wildfires, pests and diseases, and acid rain.
- These all contribute to a loss of biodiversity.

 Over to you

List the advantages and disadvantages of fire for the taiga.

You need to know:

- the advantages and disadvantages of global actions to protect rainforests.

Student Book
See pages 274–275

CITES

The Convention on International Trade in Endangered Species (CITES) is an international treaty adopted by 180 countries.

- It lists 34,000 endangered species of animals and plants.
- It bans international trade in listed species.

Advantages	Disadvantages
Many countries have signed up; countries co-operate on trade.	Protects **species**, not **ecosystems**, so does not prevent deforestation.
Protected species include a wide variety geographically and by type.	Global warming could undermine its success.
Has had some key successes, e.g. reducing the ivory trade and halting the decline of African elephants.	Relies on countries setting up and funding monitoring and policing systems, which many low-income countries (LICs) cannot afford.
Works well for high profile 'cute and cuddly' threatened species such as the snow leopard.	Species have to be under threat to get 'on the list' by which time the problem may be too serious to solve.

Figure 1 The advantages and disadvantages of CITES

REDD

Reducing Emissions from Deforestation and forest Degradation (REDD) is a United Nations project to stop deforestation, one of the main causes of climate change.

- Aims to **conserve** and enhance forest carbon stocks and sustainably manage forests.
- Governments and TNCs fund projects to conserve forests in developing countries while offsetting their own carbon emissions.
- Some argue that offsetting is an easy way for developed countries to *appear* to reduce their emissions without actually doing anything.

Six Second Summary

- Two of the main global actions to protect rainforests are CITES and REDD.
- Juma SFR is an example of a REDD project.
- Actions to protect rainforests have advantages and disadvantages.

Juma Sustainable Forest Reserve (SFR)

This is an area of pristine tropical rainforest in the Amazon.

- It was protected in 2006 as Brazil's first REDD project.
- Juma SFR (run by a local NGO) pays local families $28 per month not to cut down the forest.
- The money is donated by the Marriott Hotels (a TNC), a Brazilian bank and the regional state government.

Key

- ☐ tropical rainforest
- ▨ deforested area
- — Juma SFR boundary

Figure 2 Maps showing estimates of deforestation that would have taken place in Juma in 2030 and 2050 had there not have been any protection

Has it been successful?	On the other hand...
• Without protection, 60% of Juma's forest would have gone by 2050. • The area is large enough to support carnivores and larger primates. • Incomes have risen. • Funding has built 7 schools, trained people in sustainable farming, provided cleaner water and added solar panels to roofs. • Ecotourism is being developed to provide extra income.	• Juma SFR relies on donations, especially from Marriott hotels. If these stop, the project may stop. • Money for families is less than $1 per day. The reserve manager earns $25,000 per year. • Local people have 'signed away' their right to use forest products. • The huge area is hard to police, so illegal logging might continue.

Figure 3 The successes and challenges ahead for Juma SFR

Over to you

Create a mnemonic to help you remember the successes of, and challenges facing, Juma SFR.

Student Book
See pages 276–277

You need to know:

- how tropical rainforests can be conserved using sustainable management.

Sustainable forest management

This conserves forests by ensuring they are not used faster than the pace at which they can be renewed. Benefits include:

- **Economic:** reducing poverty by creating incomes from alternative livelihoods.
- **Social:** improving facilities to benefit the community (e.g. health centres or schools).
- **Environmental**: protecting biodiversity alongside other resources.

Kilum-Ijim: Sustainable environmental management

Kilum-Ijim is an area of mountain rainforest in Cameroon, Africa and home to 35 communities from three tribes.

- It's under pressure from farming and logging for timber and fuel.
- Local groups created a sustainable forest reserve (Figure 1) in 1987.
- They worked with communities to develop rules for the use of the forest and to educate people about replanting trees and safe levels of hunting and logging.

How successful is the project?

- There has been an 8% increase in forested area since the project began.
- People have been provided with alternative livelihoods (agroforestry and ecotourism) so that they don't have to cut down the forest for income.

Future challenges

- Population growth will increase the pressure on deforestation and bring urban areas and roads nearer to the forest.
- Support from international donors could end.
- Climate change could degrade the forest.

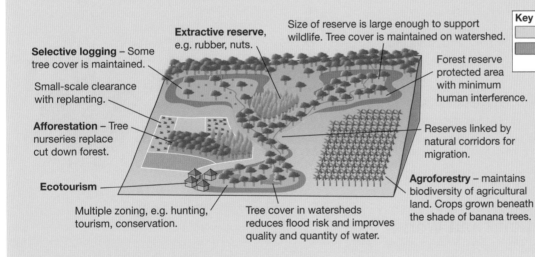

Selective logging – Some tree cover is maintained.

Extractive reserve, e.g. rubber, nuts.

Size of reserve is large enough to support wildlife. Tree cover is maintained on watershed.

Key
- Core conservation area
- Buffer zone – light use on rotational basis

Small-scale clearance with replanting.

Forest reserve protected area with minimum human interference.

Afforestation – Tree nurseries replace cut down forest.

Reserves linked by natural corridors for migration.

Ecotourism

Agroforestry – maintains biodiversity of agricultural land. Crops grown beneath the shade of banana trees.

Multiple zoning, e.g. hunting, tourism, conservation.

Tree cover in watersheds reduces flood risk and improves quality and quantity of water.

Figure 1 The sustainable forest of Kilum-Ijim in Cameroon, Africa. Lots of different techniques have been used to achieve sustainable environmental management.

Six Second Summary

- Sustainable management of tropical rainforests has economic, social and environmental benefits.
- The Kilum-Ijim project has been a success, but faces challenges in the future.

 Over to you

Create a Venn diagram with circles for 'economic', 'social' and 'environmental' impacts. Complete the diagram with the impacts of the project in Kilum-Ijim.

Student Book
See pages 278–279

You need to know:

- how taiga forests are managed and protected.

Managing taiga wilderness

There are pressures to develop the taiga for:

- oil, gas and mineral extraction, and HEP
- timber for making paper and construction.

In the USA, the **wilderness** is a protected area untouched by human activity. In these areas:

- motorised transports is not allowed
- people must leave no trace of their activities (e.g. camping)
- logging, mining and road building are banned.

National Parks

Many taiga areas are designated as national parks in the USA, Canada and Russia.

National parks usually:

- exceed 1000 hectares in size
- have legal protection
- are open to the public for recreation.

Wood Buffalo National Park, Canada

The world's second largest national park, Wood Buffalo, was created in 1922 to protect mountain bison from hunting. Because the mountain bison is so rare, Wood Buffalo became a UNESCO World Heritage Site and a RAMSAR wetland.

Even though it is protected, threats exist.

- Wood Buffalo lies north of the Athabasca tar sands mining area (Section 8.6).
- Tar sands mining is proposed close by, potentially polluting the river and reducing water flow.
- HEP dams could disrupt wetlands and river flow.

In 2015, UNESCO warned the Canadian Government that it was not doing enough to protect Wood Buffalo.

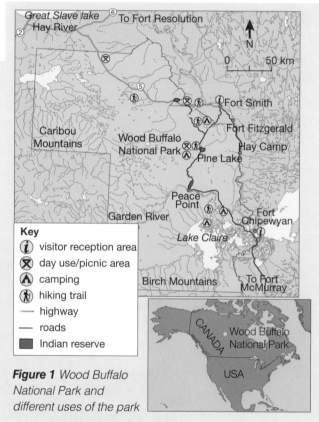

Key

- (i) visitor reception area
- (X) day use/picnic area
- (A) camping
- (hiker) hiking trail
- — highway
- — roads
- ▪ Indian reserve

Figure 1 *Wood Buffalo National Park and different uses of the park*

RAMSAR wetland and World Heritage Sites

These are two types of conservation status for areas of global importance.

- Each gives areas under threat an extra level of protection.
- Countries submit their own sites to be added to these lists. In return, they must agree to conserve these areas, and provide funding for protection.

 Six Second Summary

- There are pressures to develop the taiga.
- Wilderness areas and national parks are ways of protecting the taiga.
- RAMSAR wetland and World Heritage Sites are types of conservation.

 Over to you

Create a table of arguments **for** and **against** protecting areas of taiga from development. List the arguments in order of their significance.

You need to know:

Student Book
See pages 280–281

- the reasons for conflicting views about the future of the taiga.

Exploiting the taiga

In some areas **clear-cutting** is used.

- It is not sustainable and increases soil erosion, landslides and river bank erosion.
- It destroys mosses, lichen and other plants on the forest floor.
- Even if the taiga is replanted, the regenerated forest has lower biodiversity.

The alternative is **selective logging**, which removes only the large valuable trees and leaves some of the forest intact.

Conflict in the taiga

Some believe the taiga should be conserved; others that it should be exploited (Figure 2). National governments have to try and balance these views which can lead to conflict (Figure 3).

Sustainable Forest Management (SFM) in Finland

People in Finland respect forests and put pressure on government and industry to protect them (Figure 1).

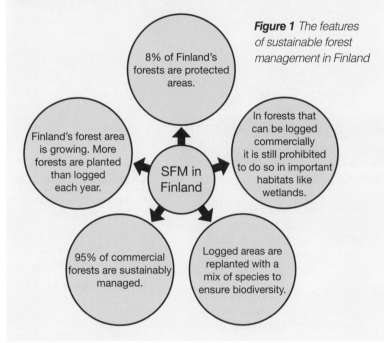

Figure 1 *The features of sustainable forest management in Finland*

- 8% of Finland's forests are protected areas.
- In forests that can be logged commercially it is still prohibited to do so in important habitats like wetlands.
- Finland's forest area is growing. More forests are planted than logged each year.
- SFM in Finland
- 95% of commercial forests are sustainably managed.
- Logged areas are replanted with a mix of species to ensure biodiversity.

World Wildlife Federation spokesperson	Athabasca Chipewyan First Nation spokesperson	Syncrude website
Conservation non-governmental organisation (NGO)	*North American indigenous group opposed to tar sands development*	*The largest oil company extracting tar sands in Canada*
'Russia's taiga zone is fragmented by roads, rail, and infrastructure developments. Coal mining, logging, pollution, oil and gas development all pose significant threats to the region.'	'Our Nation is not against development of our lands and territory - we want to see the respectful sharing and utilization of the land. Economic development at the expense of people and the planet makes no sense.'	'We have a long and proud history of contributing to the economic well-being of Canadians by providing rewarding employment to thousands of people, through the payment of taxes, and through the purchase of goods and services from suppliers.'

Figure 2 *Contrasting views about the taiga*

Six Second Summary

- Selective logging is a more sustainable forestry practice than clear-cutting.
- In Finland forests are managed sustainably.
- There is conflict in the taiga between conservation and exploitation.

Over to you

For each of the opinions given in Figure 2, write a paragraph explaining why this person or group has this viewpoint.

Indigenous people (traditional use of forest resources)					
✓	Environmental groups				
🖑	✓	National park and wilderness area managers			
🖑	✓	✓	Climate and biodiversity scientists		
🖑	🖑	🖑	✓	Visitors and tourists	
×	×	×	×	×	Oil, gas, mineral extraction companies
×	×	×	🖑	🖑	🖑	HEP companies

× Major conflict   🖑 Minor / manageable conflict   ✓ No conflict

Figure 3 *Different views about how the taiga should be managed can lead to conflict*

Topic 9
Consuming energy resources

Your exam

- Topic 9 Consuming energy resources is part of Paper 3, People and environment issues – making geographical decisions.

- Paper 3 is a 90-minute written exam and makes up 25% of your final grade. The whole paper carries 64 marks (including 4 marks for SPaG).

- You must answer all parts of Paper 3. Section A contains questions on Topic 7 People and the biosphere (pages 130–137), Section B on Topic 8 Forests under threat (pages 138–149), and Section C on Topic 9 Consuming energy resources. Section D is a decision-making question which draws together all three topics.

Tick these boxes to build a record of your revision

Your revision checklist

Spec Key Idea	Detailed content that you should know	1	2	3
9.1 Energy resources can be classified in different ways and their extraction and use has environmental consequences	• How energy resources can be classified as non-renewable, renewable and recyclable			
	• How mining and drilling can have environmental impacts, and the landscape impacts of renewable energy			
9.2 Access to energy resources is not evenly distributed which has implications for people	• How access to energy resources is affected by access to technology and physical resources			
	• The global pattern of energy use per capita and the causes of variations			
9.3 The global demand for oil is increasing, but supplies are unevenly available	• How oil reserves and production are unevenly distributed and why oil consumption is growing			
	• How oil supply and oil prices are affected by changing international relations and economic factors			
9.4 The world's continuing reliance of fossil fuels increases pressure to exploit new areas	• Economic benefits and costs of developing new conventional oil and gas sources in ecologically-sensitive and isolated areas			
	• Environmental costs of developing new unconventional oil and gas sources in ecologically-sensitive and isolated areas			
9.5 Reducing reliance on fossil fuels presents major technical challenges	• The role of energy efficiency and energy conservation in reducing demand, helping finite energy supplies last longer and reducing carbon emissions			
	• Costs and benefits of alternatives to fossil fuels and future technologies aimed at reducing carbon footprints, improving energy security and diversifying the energy mix			
9.6 Attitudes to energy and environmental issues are changing	• How different groups have contrasting views about energy futures			
	• How, in some developed countries, rising affluence, environmental concerns and education are changing attitudes to unsustainable energy consumption and reducing carbon footprints			

*Student Book
See pages 284–285*

You need to know:

- how energy resources can be classified into different categories.

How are energy resources classified?

There are three main categories:

- **Non-renewable:** Once used up, cannot be replaced; includes oil, coal and gas (**fossil fuels**); they are finite resources.
- **Renewable:** Will never run out and are infinite resources; includes wind power, solar power and hydroelectric power (HEP).
- **Recyclable:** provides energy from sources that can be recycled or reused; includes nuclear (reprocessed uranium) and biofuel energy.

Non-renewable energy: natural gas supplies in Europe

- Used for electricity production, heating and cooking.
- 60% of UK gas comes from the North Sea via a pipeline from Norway.
- Russia and Ukraine export large amounts of natural gas to Europe.
- Global gas use is predicted to peak in 2030, supplies are then likely to fall.

Recyclable energy: biogas in India

- Organic matter (e.g. wood chips, animal dung) ferments in a biogas plant and releases methane.
- The gas is collected in a tank and used to power electricity generators or for cooking.
- There are over 2.5 million biogas plants across India which provide energy for rural villages.

Figure 2 *A biogas plant, Trivandrum, India*

Renewable energy: wind power in the USA

- There are over 16 000 large wind turbines in California.
- Hundreds of homes use smaller turbines to convert wind power into electricity.
- In 2015, wind power provided 6.9% of California's total energy requirements.

Figure 1 *Wind turbines in San Gorgonio, just outside Palm Springs, California*

 Six Second Summary

- Energy resources are classified as non-renewable, renewable and recyclable.
- Natural gas is non-renewable, and global gas supplies are likely to fall after 2030.
- Wind power is an example of renewable energy.
- Biogas is a recyclable form of energy.

Over to you

Classify the following energy resources into either non-renewable, renewable, or recyclable: solar power, oil, wave power, coal, wind power, water, nuclear power, biogas, wood, biomass, HEP, natural gas, geothermal, and tidal power.

You need to know:

- how energy extraction and production affects the environment and landscapes.

Student Book
See pages 286–287

Wind turbines and solar panels – good or bad?

Wind and solar power can be used to produce electricity.

- They reduce **carbon emissions** by using a renewable resource.
- Some people argue that wind turbines and solar panels look out of place in rural areas.
- Regardless, UK energy requirements mean that large areas of the rural landscape will be used in the future for wind turbines or solar panels.

Oil spill: drilling in the Gulf Coast, USA

When offshore oil drilling goes wrong, it causes vast oil leaks.

- In 2010, BP's Deepwater Horizon oil rig exploded in the Gulf of Mexico, killing 11 people.
- Oil leaked into the sea for 87 days; a total of 3.2 million barrels was lost.
- Marine and bird life were killed as oil coated the ocean's surface.

Landscape scarring: Xilinhot, China

Coal is China's biggest energy resource, providing 70% of its energy.

- China is increasingly using surface opencast coal quarries instead of underground mines (Figure 1).
- Environmental NGOs oppose these large mines because they create scars on the landscape and use huge amounts of water to extract the coal.

Figure 1 *The Shengli opencast coal mine in Xilinhot, Inner Mongolia*

Deforestation: HEP development in Pará, Brazil

The Belo Monte Dam in in Pará, Brazil (Figure 2) will be one of the world's largest dams and will provide electricity.

- HEP provides 85% of all electricity used in Brazil.
- In order to build the HEP stations in Pará, rainforest needed to be cleared.

Figure 2 *The world's third largest hydroelectric dam will submerge 400 km² of rainforest, drowning people's homes, and wildlife habitats*

 Six Second Summary

The extraction and production of energy (from renewable sources such as solar panels and wind turbines, plus mining, oil drilling and HEP production) has impacts on the environment and landscape.

Over to you

For each type of energy included on this page, draw a table to show its environmental advantages and disadvantages.

Student Book
See pages 288–289

You need to know:

- how access to energy resources is affected by accessibility and technology.

Coal – accessibility versus technology

For 200 years, millions of tonnes of coal was mined in the UK.

- The coal formed during the Carboniferous period over 300 million years ago.
- It is the remains of giant tropical plants from swamp forests. When they died, they formed layers (seams) which geological pressures turned into coal.
- There is still coal in the UK but the remaining seams are deep and hard to access.
- Although **technology** makes mining possible, coal would be more expensive than other energy sources.

Decline of the UK coal industry

Production declined sharply during the 1970s, and had almost disappeared by 2015 due to:

- The high cost of coal mining.
- Cheaper imported coal from Russia and the USA.
- Declining demand as trains and homes switched from coal.
- Use of other energy sources (e.g. oil).
- Pressure to reduce **greenhouse gas emissions**; coal produces more greenhouse gases than any other fuel.

Meeting UK energy needs

Coal is still important, but the UK now has a more varied **energy mix**.

- **Fossil fuels** – Updated drilling technology and deep sea oil rigs meant that natural gas and oil were accessible from the North Sea. Production is now declining.
- **Renewable energy** – Wind, solar, biomass and HEP currently provide 10% of the UK's energy. Most wind power is offshore (Figure 1) but the cost of building turbines there is high.
- **Recyclable energy** – Nuclear energy provides 16% of UK energy. Hinkley Point C in Somerset will be the first in a new generation of reactors.

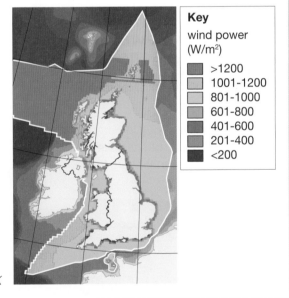

Key
wind power (W/m²)

- >1200
- 1001–1200
- 801–1000
- 601–800
- 401–600
- 201–400
- <200

Figure 1 *Wind power potential offshore in the UK*

Six Second Summary

- Access to energy resources depends on technology and accessibility.
- Coal was an important energy source for the UK but its use has declined.
- The current UK energy mix includes fossil fuels, renewable energy and recyclable energy.

Over to you

Which type of energy (fossil fuels, renewable or recyclable energy) will best enable the UK to meet its future energy needs? Explain, and justify, your choice.

Geographical skills: investigating global energy resources

*Student Book
See pages 290–291*

You need to know:

- how to use geographical skills to investigate resource patterns.

Where does the world's energy come from?

Fossil fuels are not spread evenly across the planet. Figure 1 shows the global coal, oil and gas deposits.

- Some places have an abundance (North America, the former Soviet Union, and the Middle East) while others have less (Australia).
- Coal is the most abundant resource, there is enough for another 150 years supply.

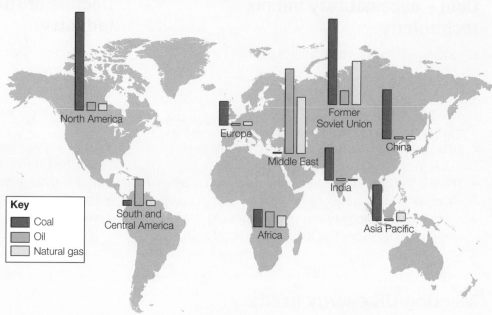

Figure 1 *Global coal, oil and gas deposits*

Identifying renewable energy potential in the UK

The UK has huge potential for renewable energy (Figure 2).

- It's the windiest country in Europe.
- It has 11 000 miles of coastline.
- It has enough daylight hours for 200 000 solar power projects.

⏱ **Six Second Summary**

- Fossil fuels are not evenly spread across the planet.
- The UK has great potential to develop renewable energy resources.

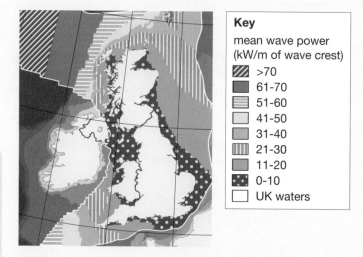

Figure 2 *The wave power potential offshore in the UK*

✏ **Over to you**

- Using the information in Sections 9.1–9.4, produce a series of mind maps to show how access to energy resources is affected by: **a)** new technology (e.g. wind turbines); **b)** geology (e.g. where coal exists); **c)** climate influences on renewables (e.g. whether or not wind or solar is possible); **d)** landscape influences on renewables (e.g. whether the infrastructure can be added to the area).
- **a)** Describe the distribution of global coal, oil and gas deposits in Figure 1. **b)** Rank the continents in terms of coal reserves. **c)** Which continent has the least potential for coal extraction?
- **a)** Describe the distribution of wave power around the UK in Figure 2. **b)** Identify two regions that would be most suitable for developing wave power. Justify your answer.

Student Book
See pages 292–293

You need to know:

- how, and why, the global pattern of energy use varies.

2014 – a global turning point?

2014 was the first year where global energy consumption looked as if it might level off. This happened because:

- cars are using less fuel
- power stations are wasting less energy
- houses are being better insulated.

Economic development and energy use

Most growth in energy consumption since 2000 came from India and China. Figure 1 shows where most energy was consumed in 2014.

- The USA has 4.5% of the world's population but uses 21% of the world's energy.
- As countries develop, energy use increases because more people own cars and domestic appliances (e.g. washing machines).

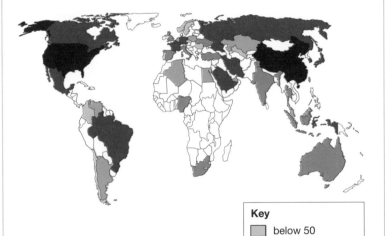

Figure 1 *Total energy consumption by country in million tonnes of oil equivalent, 2014. Notice that the highest energy users are the USA and China.*

Key

	below 50
	50-100
	100-200
	200-2000
	above 2000
	No available data

Demand from different economic sectors

As well as using *more* energy, the *types* of energy used changes as countries develop.

- Low-income countries (LICs) e.g. Malawi, have mainly primary economies. Only a small amount of energy is used, mainly in the home.
- Newly industrialising or emerging countries (e.g. India) focus on manufacturing, using large amounts of energy, often coal.
- People in high-income countries (HICs) consume more products, so demand for energy grows, often gas or oil.
- Many HICs have 'exported' manufacturing to countries such as China, making their overall energy use appear lower.

Relying on traditional fuels

Many LICs suffer from **energy poverty**. They lack electricity or money to pay for it and collect fuel (wood or animal dung) themselves.

- 20% of the world's population have no access to electricity.
- 33% of the world's population rely on biomass for cooking and heating.
- Their health suffers due to lung conditions caused by smoke from traditional wood- or dung-burning stoves.

Six Second Summary

- Global energy use has increased since 2000 but the growth is slowing.
- The more developed a country is, the more energy it uses.
- Many LICs suffer from energy poverty.

Over to you

Using the information on this page, draw a spider diagram to show the relationship between energy use and economic development.

Student Book
See pages 294–295

You need to know:

- how oil reserves and production are unevenly distributed
- where, and why, oil consumption is increasing.

Where does the world's oil come from?

Oil is used all over the world to fuel cars, heat buildings, provide electricity and make plastics. But its production is unevenly distributed (Figure 1).

Key	
world	88672.62
North America	18720.62
S & Cent. America	7613.07
Europe & Eurasia	17197.82
Middle East	28554.67
Africa	8262.91
Asia Pacific	8323.54

Figure 1 *Oil production by region, 2014, in millions of barrels per day. The Middle East is the biggest producer with North America second largest. New reserves of oil shales in the USA and Canada are changing this.*

Drilled out?

Exactly how much oil lies underground is unknown. As more is used, it's likely to become more difficult and expensive to extract.

- Some believe that **peak oil** (where half of known oil reserves have been used) has already been reached, so oil will only get more expensive as supplies fall.
- Others believe that peak oil is decades away because so much remains undiscovered.
- New reserves are discovered all the time. Estimates in 2014 showed reserves of an extra 330 billion barrels (ten years supply), compared to estimates in 2004.

Rapid industrialisation in China

- Since 1990, China's rapid industrialisation has increased the production of goods for export.
- This has caused a massive increase in energy demand – first for coal, then oil.
- China's oil consumption doubled between 2004-2014.

Six Second Summary

- Oil reserves and production are unevenly distributed.
- Oil is a finite resource that will run out one day, but it's unknown when.
- Peak oil is the point at which half of the known oil reserves have been used.
- Oil consumption continues to rise as countries (especially those in Asia) develop.

Growing demand for oil

Demand for oil in high-income countries may have peaked, but a significant increase in consumption is likely to come from emerging economies (Figure 2).

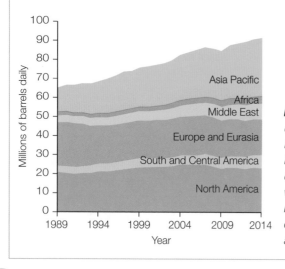

Figure 2 *Global consumption of oil by region. As Asia becomes wealthier, demand for oil will increase as more people use energy for domestic appliances and cars.*

Over to you

Draw two spider diagrams – one to show reasons why demand for oil continues to increase, and the second to show why oil is likely to become more expensive in future.

The changing price of oil

Student Book
See pages 296–297

You need to know:

- how oil supply and oil prices are affected by changing international relations and economic factors.

Oil – a licence to print money?

Oil prices change regularly. Between 2005-2010 oil prices rose for a number of **economic** and **political** reasons that caused **supply shortages**.

- Nigeria had security problems that shut down 20% of its oil production.
- In Mexico, oil companies cut production as a protest against political interference.
- Conflict between Russia and Ukraine in 2008 meant Ukraine refused to pay increases in Russian gas prices. (Gas and oil prices work together – as one rises, so does the other.)
- OPEC members (led by Saudi Arabia) restricted oil supplies to keep prices high.

Year	Average price per barrel (US$)
1975	12.21
1980	36.83
1985	27.56
1990	23.73
1995	17.02
2000	28.5
2005	54.52
2010	79.5
2015	55.0

Figure 1 The average global price of one barrel of crude oil, 1975-2015. Notice how it fluctuates over time.

The Iraq War 2003-2011

In 2003, US and Allied forces invaded Iraq.

- Iraq has the world's 4th largest oil reserves.
- Its then leader Saddam Hussein was thought to pose a threat to global oil supplies.
- The conflict led to shortages of oil and increased prices.
- Other Allies (e.g. Saudi Arabia) then increased production to stabilise prices.

The era of fracking

In recent years, it has become viable to extract oil from shale rocks.

- Water is blasted into rock fractures under pressure to release the oil. This process is called hydraulic fracturing or fracking.
- It has made the USA less reliant on imported oil because it has large shale deposits.

Why do oil prices change?

The global oil price is based on several factors.

- **Demand**. High demand causes prices to rise. Falling demand causes lower prices.
- **Supply**. If there is too much oil the price falls, too little and it rises.
- **Political decisions**. Countries sometimes increase supply to increase income.
- **New supplies**. In 2015 Saudi Arabia, Iraq, Iran and fracking in the USA caused the oversupply of oil and prices fell.

Six Second Summary

- Oil prices can vary for economic and political reasons.
- The Iraq war was closely linked to global oil supplies.
- Extracting oil from shale rock through 'fracking' has made the USA less dependent on oil imports.

Over to you

Explain the impact the 2003 Invasion of Iraq had on the global price of oil and why.

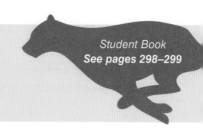

*Student Book
See pages 298–299*

You need to know:

- the economic benefits and environmental costs of developing oil and gas in environmentally sensitive areas.

Gazprom – the polar pioneers

In 2014, Gazprom (one of Russia's largest oil and gas companies) began drilling for oil in the remote northwest of Russia, within the Arctic Circle. Drilling here presents challenges.

- US$4 billion was spent developing technology to search for oil and gas beneath the sea.
- A giant oil rig was built to drill through rock in this cold, fragile environment.
- Greenpeace International protested against the drilling.

Exploring remote places

High oil and gas prices (and profits) throughout the 2000s allowed oil companies to drill in regions that were previously too expensive or difficult to work in.

New technologies enable exploitation in extreme environments, where returns are high.

- **Drilling technology** allows access to deep-water reserves.
- **Seismic imaging** detects rock structures containing oil and gas.
- **Liquefaction** of natural gas (converting gas into liquid) makes it possible to transport gas by ship.

The Ichthys LNG project

- The Ichthys gas field, one of the world's largest, lies 220 km off the coast of north-western Australia.
- It is 260 metres below sea level and produces 100 million tonnes of liquefied natural gas (LNG) per day.
- 900 jobs were created in construction.

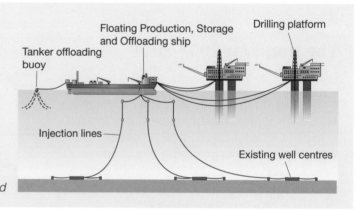

Figure 1 How liquid natural gas is obtained

The Arctic at risk?

The Arctic has up to 25% of the world's remaining oil and gas.

- It's a huge area of wilderness. The region's taiga forest is 25% of all remaining forest globally.
- Several countries claim the Arctic (Figure 2).
- Environmental groups worry that oil companies will cause more damage to Alaska and Siberia.

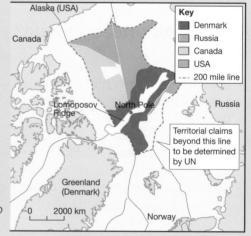

Figure 2 Territorial claims in the Arctic. This is likely to cause conflict in future over access to oil and gas.

Six Second Summary

- High profits and new technology enabled oil and gas companies to drill in regions previously too expensive or difficult to access.
- The Arctic is at risk from the exploitation of oil and gas reserves.

 Over to you

Draw a table of the environmental costs and economic benefits of developing oil and gas reserves in the Arctic.

The costs of developing fossil fuels 2

*Student Book
See pages 300–301*

You need to know:

- about the environmental costs of developing unconventional oil and gas resources in ecologically sensitive and remote areas.

Unconventional fossil fuels

Using tar sands to produce oil and gas is **unconventional** – i.e. it is different from how they are usually produced. It is expensive, and is possible only with high oil and gas prices and improved technology.

- Vast reserves of oil and gas are stored in areas such as Athabasca in Canada's central wetlands (Figure 1).
- Extracting these has a huge environmental cost.

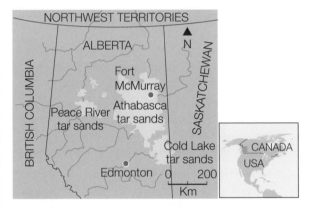

Figure 1 *The location of Canada's tar sands*

Extracting tar sands and shale gas

- **Tar sands** are naturally-occurring mixtures of sand, clay, water and **bitumen** (a form of petroleum).
- Bitumen is extracted from tar sands by injecting steam underground. It uses enormous amounts of energy and water.
- **Shale gas** is found trapped underground in shale rock.
- Shale is impermeable, so fracking is used to extract the gas.

Tar sands oil extraction in Athabasca, Canada

This area is mainly home to taiga forest and peat bogs.

- Large-scale open pit mining to extract oil requires the destruction of forests and peat bogs, causing a massive loss of ecosystems (Figure 2).
- Greenpeace is concerned that tar sands oil extraction has led to a decline in caribou, lynx and wolverines.
- It takes 2-5 barrels of water to produce 1 barrel of oil.
- There have been reported leaks of water polluted with oil into nearby rivers.

Figure 2 *The impact of Canada's tar sands oil extraction in Athabasca. Notice the deforestation and loss of habitats that will have occurred.*

 Six Second Summary

- Shale gas and tar sands oil extraction (unconventional fossil fuels) is only possible because of technological improvements, and high energy prices.
- There are many environmental concerns about tar sands oil extraction, including the use of huge amounts of energy and water.
- Canada is a key player in the extraction of oil from tar sands.

 Over to you

Complete a table to show the costs and benefits of developing tar sands and shale gas.

Student Book
See pages 302–303

You need to know:

- how energy efficiency and conservation can cut demand, help finite energy supplies last longer, and reduce emissions.

How much longer can it last?

Fossil fuel use contributes to the world's **carbon footprint.**

- Fossil fuels release greenhouse gases when burnt, increasing our carbon footprint and contributing to climate change.
- Fossil fuels are finite; reducing their use means they'll last longer and also cut carbon emissions.

Energy efficiency and conservation at home

There are several ways to reduce energy consumption at home (Figure 1). To help achieve this the UK Government offered a 'Green Deal'.

- Loans were offered, whose repayments were added to electricity bills.
- Grants up to £1250 were offered towards the cost of installing two energy-saving home improvements.

The government stopped funding the Green Deal in 2015 and subsidies offered to people to install wind turbines and solar panels ended in 2012.

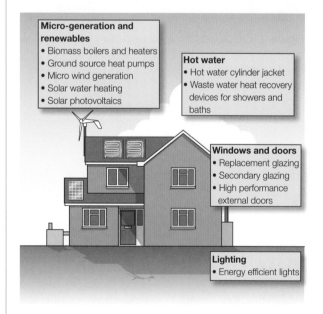

Micro-generation and renewables
- Biomass boilers and heaters
- Ground source heat pumps
- Micro wind generation
- Solar water heating
- Solar photovoltaics

Hot water
- Hot water cylinder jacket
- Waste water heat recovery devices for showers and baths

Windows and doors
- Replacement glazing
- Secondary glazing
- High performance external doors

Lighting
- Energy efficient lights

Figure 1 Improvements to help energy efficiency and conservation

Improving transport energy efficiency

Vehicle Excise Duty (VED), or road tax, depends on the level of vehicle pollution.

- Vehicles with low CO_2 emissions pay no road tax to encourage drivers to buy them.
- Electric cars have low emissions but a relatively low range before needing recharging.

Reducing London's carbon footprint

- All new London buses have been hybrid since 2012 (Figure 2). They are 40% more fuel-efficient and produce 40% less CO_2.
- The Santander cycle hire scheme was introduced in 2010 with 6000 bikes.
- 12 new cycle superhighways (with wide, blue surfaces) were built by 2015.

Figure 2 A hybrid bus in London

Six Second Summary

- The global carbon footprint is rising.
- Energy efficiency and conservation measures can cut the amount of energy used at home.
- Transport technology and initiatives in large cities like London can play a part in reducing the amount of energy used.

Over to you

Write a list of four actions each that **a)** people and **b)** companies could take to reduce their carbon footprint.

You need to know:

- the costs and benefits of alternatives to fossil fuels.

Student Book
See pages 304–305

Will the lights go out?

In the future, the world may lack **energy security**. Fossil fuels generate 82% of global energy supplies but they are a finite resource.

- Countries need to **diversify** their energy mix and use different sources.
- As global energy demands grow, renewable energy will become more important to reduce carbon footprints and ensure energy security.

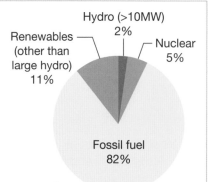

Hydro (>10MW) 2%

Renewables (other than large hydro) 11%

Nuclear 5%

Fossil fuel 82%

Figure 1 *The current sources of global energy supplies*

Diversifying the energy mix

Hydroelectric Power (HEP) in California

- It uses water to drive a turbine that generates electricity.
- It is a major energy source in California, providing 12% of its electricity in 2013.
- Constructing dams to provide the water is expensive.

Solar energy in California

- Concentrating solar power (CSP) units generate electricity by using the sun's energy to heat a fluid and produce steam (which powers a generator).
- Solar photovoltaic (PV) directly coverts solar energy into electricity using PV cells grouped into panels.
- California's warm climate and sunlight is ideal for solar generation.
- It has 400 000 small-scale solar projects and the world's largest solar thermal power plant.

Biofuels

- These are renewable fuels, produced from organic matter e.g. straw, wood chippings or animal waste (biomass).
- Bioethanol is the most common; it's blended with petrol for cars.
- Biofuels have lower carbon emissions than fossil fuels, though cost the same as oil to produce.

Hydrogen technology and Toyota

- Once separated from other elements hydrogen can provide an alternative to oil.
- Separating hydrogen requires energy; renewable sources can provide this.

Figure 2 *A Toyota Mirari uses hydrogen in a fuel cell which converts hydrogen into electricity to power the car's four motors. It has a range of 312 miles before refuelling is required. The exhaust only produces water.*

Six Second Summary

- Meeting the global energy demand and ensuring energy security, is a challenge.
- Energy diversification including the use of HEP, biofuels, solar energy and hydrogen technology are ways to meet the challenge.

Over to you

Compare the advantages and disadvantages of the ways to diversify the energy mix included on this page.

Student Book
See pages 306–307

You need to know:

- that different groups have different views about energy futures.

World energy outlook

In 2015, the *World Energy Outlook* report predicted that:

- fossil fuel prices will remain unpredictable
- they will dominate until 2030, contributing to climate change.

There are two possibilities for the future of global energy use:

- **Business as usual** – assumes that the world will continue to rely on fossil fuels. Oil and gas production will increase to meet demand.
- **A sustainable future** – renewables are adopted to reduce CO_2 emissions as a way of combating impacts of climate change.

Different economic and political factors, and key players, will determine what happens. This includes who wins in discussions which involve people who:

- have vested interests in energy production
- are concerned with the impacts of climate change.

The 450 Scenario

The International Energy Agency (IEA) set out a proposal to limit greenhouse gases in the atmosphere to 450 parts per million of CO_2.

- It would reduce the increase of global temperatures from 6°C to 2°C.
- It assumes different countries will adopt emissions targets.
- Countries would add a **carbon tax** to the use of fossil fuels.

Energy futures – views of the different players

Player	What they say
Shell, a TNC	Shell say they are committed to: • finding ways to provide energy from cleaner sources • helping customers to use energy more efficiently.
Department of Energy and Climate Change, UK Government	Committed to: • tackling climate change • keeping energy bills low for customers. Believes that: • solutions to climate change will come through innovation, technology, enterprise and competition • cutting greenhouse gas emissions should be drastic now, as any delay will mean worse cuts in future.
Climate scientists	Some believe that: • climate risks are serious and should be minimised • the world has huge development needs • the world's current energy mix is not sustainable • engineering the planet to cope with climate change is very dubious.
Greenpeace, an environmental NGO	Proposes an 'Energy Revolution' which would protect the climate by: • phasing out fossil fuels • investing in renewable energy.
Consumers	Would like their energy bills kept as low as possible, yet are increasingly concerned with the threat of climate change and the impacts this will have.

Figure 1 *Energy futures – different groups have different views*

Six Second Summary

- Fossil fuels are expected to dominate the energy mix until at least 2030.
- There are two possibilities for the future of global energy use: 'business as usual' and 'a sustainable future'.
- Different groups have different views about energy futures.

Over to you

For each of the players named in the table, list two reasons why they have adopted their viewpoint.

Student Book
See pages 308–309

You need to know:

- how attitudes are changing towards energy consumption, and ecological footprints.

Changing attitudes to the environment

Environmental concerns have become more important politically.

- In the EU, green political parties and pressure groups (e.g. Greenpeace) have emerged.
- **Kuznets Curve** (Figure 1) shows the point at which development, water and food supply are no longer priorities because they have been achieved so people realise that pollution and environmental damage are problems.

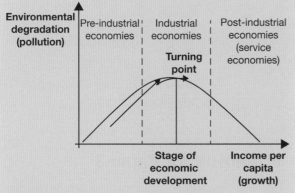

Figure 1 *The Kuznets Curve*

Educating for a changing world

Education and action are essential to the future of the planet. Schools:

- increasingly teach about sustainability, focusing on local actions people can take
- can change attitudes to climate change. (e.g. promote recycling).

Rising affluence and the environment

Affluence (rising wealth) places greater pressure on the planet.

- **Food miles** increase as people demand more exotic foods.
- But people in developed countries also 'care more' about the environment and resources with rising affluence.
- The world's most affluent countries have started to reduce energy consumption per capita by using technology (e.g. LED bulbs).

Ecological footprints

Ecological footprint can be calculated from a measure of **carbon footprint**.

- Knowing our ecological footprint allows us to reduce it and move towards **sustainable development**.
- An ecological footprint of 1 means that people live within the Earth's ability to supply resources.
- The higher the figure, the more space is needed to supply resources.

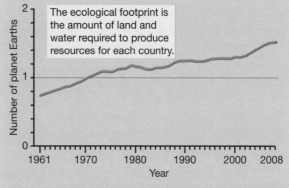

Figure 2 *Humanity's ever-increasing ecological footprint*

Six Second Summary

- Rising global affluence initially means greater global energy use per capita.
- At a point affluent countries begin to care more about the environment.
- The global ecological footprint is increasing.
- Education and action are needed to achieve sustainable development.

Over to you

Explain the differences between carbon footprint, ecological footprint and sustainable development.

Student Book
See pages 310–311

You need to know:

- how to prepare for Paper 3, the decision-making examination (DME).

What's different about Paper 3?

Paper 3 is very different from Papers 1 and 2.

- The paper is a decision-making exam about a geographical issue, linked to Topics 7, 8 and 9 in the specification (Figure 1). The issue will be about the biosphere (Topic 7) and resources (Topic 9), focusing on a place which will be located in either a Taiga or Tropical rainforest region (Topic 8).
- The issue will be outlined in a Resource Booklet, about 10 pages long. It will be about a place that you have probably not studied. Don't worry about this, as what's being assessed is your ability to read the booklet and interpret it. It's not about your knowledge of that place.
- The exam lasts for 1 hour 30 minutes. It has a total of 64 marks, including four for spelling, punctuation, grammar (SPaG). This means it's less pressured for time than the other papers.
- The last question in the exam will ask you to make a decision about the issue. That will need some thinking and planning time – that's why the timing of the paper is different.

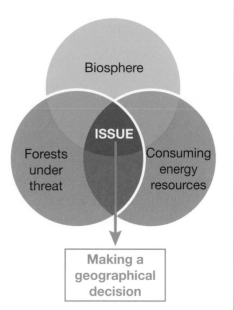

Figure 1 *How Topics 7, 8 and 9 link together*

How to spend time in the exam

- It's essential that you read the Resource Booklet carefully. This will help you become familiar with the information in it.
- The Resource Booklet will give you information about the place, and lead up to proposals for the future. It will consist of a mix of text, maps, photographs, graphs, tables of data, and views and opinions about the issue.
- Use these resources to help you make sense of the place and the issue.
- Most of the exam questions will test your understanding of the booklet, though some will test your wider understanding of Topics 7, 8 and 9.
- Some of the questions will involve making calculations, so you'll have access to a calculator.

Preparing for Paper 3

Both the Resource Booklet and exam paper are organised into sections:

- Section A: People and the biosphere (Topic 7)
- Section B: Forests under threat (Topic 8)
- Section C: Consuming energy resources (Topic 9)
- Section D: Making a geographical decision (Topics 7, 8 and 9)

To prepare for the exam, you will need to revise all three topics because some of the shorter questions will test your knowledge and understanding of them. You will also need to apply your knowledge and understanding to the issue in the Resource Booklet.

You need to know:

- about the threats Ecuador's rainforest faces from oil exploitation.

Section A: People and the biosphere

The issue: Will any economic benefits gained from exploiting the rainforest justify the damage caused?

- Oil and timber companies have recently opened up rainforests in eastern Ecuador.
- Their activity is having an impact upon traditional lifestyles, and the rainforest itself.
- Oil accounts for half of Ecuador's exports, earning about US$11 billion a year. But many believe this revenue has only benefitted a minority of Ecuador's 16 million people.
- Ecuador has decisions to make about the future of its rainforest, and the implications of developing its resources for its people, environment, and the economy.

Introduction

- Global demand for oil continues to increase (see Figure 1), as with other fossil fuels.
- To meet global demand, oil companies have searched in areas of rainforest, as traditional sources of oil (e.g. the North Sea) are used up.
- Oil was discovered in eastern Ecuador (see Figure 2) during the 1960's. It has

been developed on a small scale compared to the Middle East, for example, but the impacts of development are enormous.

- Critics argue that hundreds of thousands of indigenous people have been affected by the impacts of oil development on their communities and on the rainforest, which has been cleared.
- Those who support the exploitation of oil in Ecuador argue that oil is needed a) to earn overseas income by exporting it, and b) to save on imported oil from overseas.

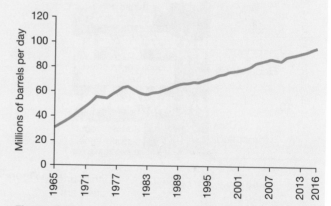

Figure 1 *Global oil consumption in millions of barrels of oil per day, 1965–2016*

Section B: Forests under threat

Figure 2 *Map of Ecuador*

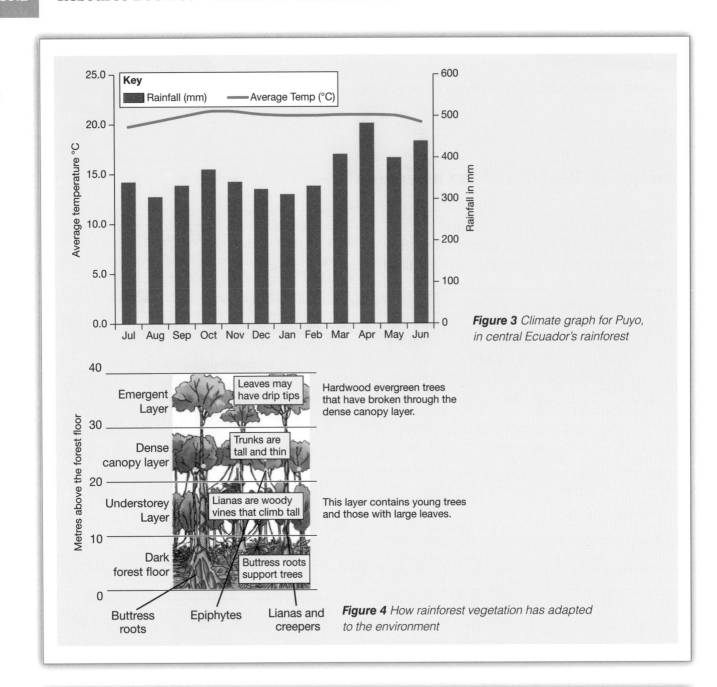

Figure 3 *Climate graph for Puyo, in central Ecuador's rainforest*

Figure 4 *How rainforest vegetation has adapted to the environment*

Section C: Consuming energy resources

Data	Ecuador	Brazil
GDP	US$ 188.5 billion (in PPP)	US$ 3.219 trillion (in PPP)
GDP per capita	US$ 11 200 (in PPP)	US$ 15 500 (in PPP)
Employment by sector	Agriculture: 28% Industry: 18% Services: 54%	Agriculture: 10% Industry: 39.8% Services: 50.2%
Population living below poverty line	26%	3.7%
Exports	Value US$ 18.3 billion Main exports: crude petroleum (29%), bananas (17.5%), cut flowers, shrimp, cacao, coffee, wood, fish	Value US$ 215.4 billion Main exports: transport equipment, iron ore, soybeans, footwear, coffee, cars
Imports	Value US$ 16.8 billion Main imports: industrial materials, fuels and lubricants, nondurable consumer goods	Value US$ 151.5 billion Main imports: machinery, electrical and transport equipment, chemical products, oil, vehicle parts, electronics

Figure 5 *Trade and economic data in 2017 comparing Ecuador (a developing economy) with Brazil (an emerging economy)*

Figure 6 *Oil exploitation by Texaco in Ecuador's rainforest*

Oil pollution and human health

The Business and Human Rights Resource Centre reported in 2010 that:

- Oil pollution in Ecuador has been substantial.
- Soil and water used by indigenous people for farming, fishing, bathing and drinking has been contaminated.

The Rainforest Action Network reported increases in serious illness; over 1400 people dying from cancers caused by pollution; high rates of birth defects; child leukaemia.

Biodiversity or cash?

- Ecuador has some of the world's most biodiverse rainforest ecosystems, e.g. Yasuni National Park in eastern Ecuador. One hectare can support 655 species of tree, 500 of fish and 600 of birds.
- In the early 2000s, Ecuador asked other countries to contribute to a campaign to help it avoid extracting oil in Yasuni National Park.
- In 2008, Ecuador was unable to repay US$ 3 billion in loans so it struck a deal with China - oil for cash.
- In 2016, Ecuador began transporting the first barrels of crude oil out of Yasuní.

Impacts on indigenous people

- Ecuador is drilling for oil in an area of rainforest inhabited by two of the last tribes in the world living in voluntary isolation.
- Several oil platforms have been established in Yasuni National Park.
- The government has ordered oil companies to leave local peoples untouched.
- But oil exploitation has led to pollution, forest clearance and the decimation of the nomadic Tagaeri and the Taromenane tribes.
- Oil companies rarely hire local people; migrant labour benefits most.

Figure 7 *Some of the impacts of oil exploitation in Ecuador*

Figure 8 *Oil spill in tropical rainforest, Ecuador*

We thought oil was going to transform our economy. I remember an article in 1966 that said Ecuador was going to be the new success story because we'd discovered oil. We thought petro-dollars were going to solve all our problems.

In the 60s there were no roads. After the oil boom in the 70s many roads were built into the Amazon and that was the key to open the door to massive development. Then the rest started. You lose biodiversity, the companies polluted water and the forest itself, affecting the people. Deforestation was so fast.

Indigenous communities depend on resources produced by tropical forests, like food, medicine, shelter. And they know they need to keep the forest to keep those renewable sources for their own lives. But the oil companies just think of the forest as a resource that they can get a lot of money from, sell it, buy more, and sell it.

Land is the main thing that traditional communities need to survive as a group. If you don't have it, you lose all the possibility to keep yourself as a human being. So their lives were changed, they lose their language, customs, culture, and territory.

The loss of biodiversity means people don't get access to the resources they need for their lifestyle, and when you don't have the forest, or you don't have the territory, you lose part of your life.

All these waste pools are still active and contain toxic materials of crude and waste water that were removed from the wells. There were several pools – three, four or five for every well that they used to dump all the toxic water produced by the wells into without any treatment.

Figure 9 *Six views from members of indigenous communities about oil exploitation in Ecuador*

Section A: People and the biosphere (8 marks in total)

Use Section A of the Resource Booklet to answer this question.

1a Define the term 'fossil fuels'. (1 mark)

1b Explain one reason why fossil fuels threaten the biosphere. (2 marks)

1ci Study Figure 1. Identify which one of the following shows global oil consumption in 1983 in millions of barrels of oil per day: (1 mark)
- 58 • 68 • 78 • 88

1cii Study Figure 1. Identify by **how much** global oil consumption increased between 1983 and 2016. Show your working. (2 marks)

1d Explain one reason why global oil consumption is likely to continue to rise steadily. (2 marks)

Section B: Consuming energy resources (8 marks in total)

Use Section B of the Resource Booklet to answer this question.

2a Study Figure 2. Identify the following features:
i Country X **ii** Line of latitude Y. (2 marks)

2b Study Figure 3. Describe two features of a tropical rainforest climate as shown in Figure 3. (2 marks)

2c Study Figure 4. Explain two ways that plants in the rainforest have adapted to the climate. (4 marks)

Section C: Forests under threat (32 marks in total)

Use Section C of the Resource Booklet to answer this question.

3a Study Figure 5. Calculate the value of crude petroleum from Ecuador's total GDP. Show your workings. Give your answer to two decimal places. (2 marks)

3b Explain two pieces of evidence from Figure 5 that show that Ecuador is not yet an **emerging** country, compared to Brazil. (4 marks)

3c Explain how increased oil production could benefit the economy of Ecuador. (4 marks)

3d Suggest one reason why some people in Ecuador believe that oil production could create problems for the country's economy. (2 marks)

3e Suggest two ways in which changes in the global oil price might affect Ecuador's oil production in the future. (4 marks)

3f Study Figures 6 and 7. Using evidence from these, assess the statement that 'oil can be as much a curse as a blessing'. (8 marks)

3g Using evidence from Figures 8 and 9, assess the extent to which the lives of indigenous peoples in Ecuador have benefited from oil exploitation. (8 marks)

Section D: Making a geographical decision (12 marks in total plus 4 marks for SPaG)

Use all the resources in Sections A to C to help you answer this question.

4 Study the three options about how Ecuador should manage its oil resources and the environment in future.

Option 1	Encourage the expansion of oil development in the rainforest to help develop the country's economy without any restrictions.
Option 2	Continue with current oil development, but do not allow any further exploitation.
Option 3	Impose heavy restrictions on all current oil developments to reduce pollution and deforestation, and convert all remaining rainforest into national parks managed by indigenous peoples.

Select the option that you think would best meet the needs of people and the forest environment in Ecuador. Justify your choice.

Use information from the Resource Booklet and from the rest of your course to support your answer. (12 marks plus 4 marks for Spelling, Punctuation and Grammar)

Section A: People and the biosphere (8 marks in total)

1a A natural fuel found underground. (1)
Within sedimentary rock e.g. oil, coal, gas. (1)
Fossilised plant material that burns to create energy. (1)

1b 1 mark awarded for correct point plus 1 for extension, e.g.:
Burning fossil fuels give off greenhouse gases (1) which causes climate change (1).

1ci 58 (1 mark)

1cii 1 mark for the answer, 1 for the working.
Any number between 95–98 millions of barrels of oil per day can be accepted as the 2016 figure (1 mark).
The subtraction gives (depending on the 2016 figure) between 37–40 millions of barrels of oil per day. (1 mark)

1d 1 mark awarded for correct point plus 1 for extension, e.g.:
Linked to increased economic development (1) which increases demand for energy (1).
Living standards increase in most countries (1) and involves buying cars / appliances which consume energy (1).

Section B: Consuming energy resources (8 marks in total)

2ai Brazil (1 mark)
2aii Equator (1 mark)

2b 1 mark awarded for correct point plus 1 for extension, e.g.:
High total rainfall (1) plus 1 mark if illustrated with total (1).
High rainfall in every month (1) plus 1 mark for variation (1).
Constant high temperatures (1) with small annual range (1).

2c Points made can either be related to the diagram or to previous knowledge and understanding. 2 x 2 marks awarded for two extended points. In each case, 1 mark awarded for correct point plus 1 for extension, e.g.:
Emergent trees grow high above the rest (1) to overcome competition for light (1).
Trees and shrubs in lower levels have to tolerate lower levels of light (1) because it's so shady (1).
Leaves of many plants have drip tips (1) to allow rainfall to drain easily (1).
Trees have buttress roots for extra support (1) because soils are thin / roots are shallow (1).

Section C: Forests under threat (32 marks in total)

3a 1 mark for the answer, 1 for the working. Correct answer US$ 54.67 billion (1).
Working e.g. 188.5 x 0.29 (1).
'188.5 x 29%' not credited as it does not show working.

3b 2 x 2 marks awarded for two extended points.
1 mark awarded for each correct point plus 1 for extension, e.g.:
Low total GDP (1) with manipulated data to illustrate e.g. '17 times greater in Brazil' (1).
Much higher employment in industry in Brazil / lower in Ecuador (1) because emerging economies have high manufacturing sector / developing countries have more in agriculture (1).
Ecuador's exports all primary products (1) compared to Brazil which exports equipment / cars / more manufactured products (1).

3c 4 x 1 marks awarded for extended points or chain of reasoning, e.g.:
Increased revenues / taxes to the government (1) which would increase government spending (1) on health or education (one example only) (1) and improve quality of life for people (1).
Increased employment in Ecuador (1) leading to a multiplier effect (1) as workers spend more on goods and services (1) and create growth in the economy (1).

3d 1 mark awarded for correct point plus 1 for extension, e.g.:
Oil is finite (1) and so might run out one day (1).
Pollution incidents might occur (1) which would cost a lot to clean up (1).

3e 2 x 2 marks awarded for two extended points.
1 mark awarded for each correct point plus 1 for extension, e.g.:
Supplies might decrease (1) which might increase the oil price / increase revenue for the government (1).
Demand might fall (1) which would cause a big fall in exports / government revenue / cuts in spending (1).

3f The mark balance is AO3 (4 marks)/AO4 (4 marks), but is awarded using the level descriptions in the table below.

- Answers should focus on the extent to which oil is an environmental curse.
- Strong candidates will recognise that the impacts of oil are only as strong as government regulation in reducing the impacts of oil exploration or of pollution.

For AO3 (Application) the arguments might include:

- Environmental quality / biodiversity of eastern Ecuador ranks among the world's highest.
- The need for legislation in controlling impacts e.g. pollution of water courses etc.
- Environmental impacts of oil exploration can be severe locally.
- Environmental damage may have consequences exceeding short-term economic benefits.
- Local pollution impacts have a knock-on effect on water pollution into the Amazon.

For AO4 (Geographical skills) information from the Resource Booklet might include:

- Evidence from Figure 6 to demonstrate forest clearance for access and drilling.
- Scale of extent of scarring, size of trucks, piles of overburden and toxic tailings.
- Possibility of water/air contamination leading to damage to fish, impact on human health.
- Any points from Figure 7 to illustrate environmental impacts.

3g The mark balance is AO3 (4 marks)/AO4 (4 marks), but marks are awarded using the level descriptions in the table below.

- Answers should focus on the extent to which indigenous peoples have benefited.
- Strong candidates will recognise that the viewpoints differ and can be categorised into e.g. economic, social and environmental.

For AO3 (Application) the arguments might include:

- Some economic benefits e.g. migrant labour.
- The cultural impacts on indigenous peoples who have shunned contact with outsiders and the threat to their lifestyle.
- Temptation by the government to override the interests of indigenous peoples.
- The environmental impacts of forest destruction, of sustainable and traditional food gathering, of pollution of water courses.

For AO4 (Geographical skills) information from the Resource Booklet might include:

- Evidence from Figure 9 from different speakers.
- The categorisation of viewpoints into economic interests, cultural and social interests and environmental impacts (with additional reference to Figure 8 or Figure 6).
- Any observations from Figure 8 to illustrate environmental impacts.

Level	Mark	Descriptor
0	0	No acceptable response
1	1–3	• Attempts to apply understanding to deconstruct information but understanding and connections are flawed. An unbalanced or incomplete argument that provides limited synthesis of understanding. Judgements are supported by limited evidence. (AO3) • Uses some geographical skills to obtain information with limited relevance and accuracy, which supports few aspects of the argument. (AO4)
2	4–6	• Applies understanding to deconstruct information and provide some logical connections between concepts. An imbalanced argument that synthesises mostly relevant understanding but not entirely coherently, leading to judgements that are supported by evidence occasionally. (AO3) • Uses geographical skills to obtain accurate information that supports some aspects of the argument. (AO4)
3	7–8	• Applies understanding to deconstruct information and provide logical connections between concepts throughout. A balanced, well-developed argument that synthesises relevant understanding coherently, leading to judgements that are supported by evidence throughout. (AO3) • Uses geographical skills to obtain accurate information that supports all aspects of the argument. (AO4)

Section D: Making a geographical decision (12 marks in total plus 4 marks for SPaG)

4   The mark balance is AO2 (4 marks), AO3 (4 marks) and AO4 (4 marks), but marks are awarded using the level descriptions.

- A case can be made for any of the three options.
- Answers should focus on the extent to which the choice of option is justified.
- Strong candidates will use prior learning as well as what is contained in the Resource Booklet.

For AO2 (Understanding) the arguments might include:

- Recognising the ecological value of the rainforest.
- Understanding why there are conflicting views on protection versus exploitation of rainforests for their resources.
- The environmental impacts on landscape, water courses, deforestation of extracting oil.
- The costs on the culture of indigenous peoples.
- The factors influencing global supply and demand and world oil prices and how these can lead to boom and bust within the oil industry.

For AO3 (Application) the arguments might include:

- Exploitation of oil has different impacts in the short- and long-term.
- People have different powers to influence development (e.g. indigenous peoples versus government and oil companies).
- Weighing up short-term gains from oil exploitation versus long-term damage to rainforests.
- The dominance of oil in Ecuador's economy and the need to diversify.

For AO4 (Geographical skills) information from the Resource Booklet might include:

- Evidence from Figures 6–9 from different speakers, interests and environmental impacts.
- The categorisation of viewpoints into economic, cultural and social interests and environmental impacts (with additional reference to Figure 8 or Figure 6).
- Any points from any other resources in the economy (e.g. Figure 5) to illustrate social, economic, political or environmental impacts.

In awarding a mark, the level descriptions below are used.

Level	Mark	Descriptor
0	0	No acceptable response
1	1–4	• Demonstrates isolated elements of understanding of concepts and the interrelationship between places, environments and processes. (AO2) • Attempts to apply understanding to deconstruct information but understanding and connections are flawed. An unbalanced or incomplete argument that provides limited synthesis of understanding. Judgements that are supported by limited evidence. (AO3) • Uses some geographical skills to obtain information with limited relevance and accuracy, which supports few aspects of the argument. (AO4)
2	5–8	• Demonstrates elements of understanding of concepts and the interrelationship between places, environments and processes. (AO2) • Applies understanding to deconstruct information and provide some logical connections between concepts. An imbalanced argument that synthesises mostly relevant understanding, but not entirely coherently, leading to judgements that are supported by evidence occasionally. (AO3) • Uses geographical skills to obtain accurate information that supports some aspects of the argument. (AO4)
3	9–12	• Demonstrates accurate understanding of concepts and the interrelationship between places, environments and processes. (AO2) • Applies understanding to deconstruct information and provide logical connections between concepts throughout. A balanced, well-developed argument that synthesises relevant understanding coherently leading to judgements that are supported by evidence throughout. (AO3) • Uses geographical skills to obtain accurate information that supports all aspects of the argument. (AO4)

Marks for SPaG

Level	Mark	Descriptor
0	0	• Writes nothing or in a style which does not link to the question, or make sense of the question.
1	1	• Spelling and punctuation reasonably accurate. Some meaning overall. A limited range of specialist terms.
2	2–3	• Spelling and punctuation show considerable accuracy. Grammar shows general control of meaning overall with a good range of specialist terms.
3	4	• Spelling and punctuation show consistent accuracy. Grammar shows effective control of meaning overall with a wide range of specialist terms.

Glossary

This glossary includes key words and terms, but it is not a prescribed list from the exam board – the final exam may include words not on this list.

*cross reference

A

abiotic non-living part of a *biome, includes the *atmosphere, water, rock and soil

abrasion the scratching and scraping of a river bed and banks by the stones and sand in the river

aftershocks follow an earthquake as the fault 'settles' into its new position

alluvium all deposits laid down by rivers, especially in times of flood

altitudinal zonation is the change in *ecosystems at different altitudes, caused by alterations in temperature, precipitation, sunlight and soil type

antecedent rainfall the amount of moisture already in the ground before a rainstorm

asthenosphere part of the Earth's *mantle. It is a hot, semi-molten layer that lies beneath the *tectonic plates

atmosphere the layer of gases above the Earth's surface

attrition the wearing away of particles of debris by the action of other particles, such as river or beach pebbles

B

bankful the *discharge or contents of the river which is just contained within its banks. This is when the speed, or *velocity, of the river is at its greatest

bar an accumulation of *sediment that grows across the mouth of a bay, caused by longshore drift

basalt a dark-coloured volcanic rock. Molten basalt spreads rapidly and is widespread. About 70% of the Earth's surface is covered in basalt *lava flows

birth rate number of live births per 1000 people per year

biodiversity means the number of different plant and animal species in an area

biofuels any kind of fuel made from living things, or from the waste they produce

biogas a gas produced by the breakdown of organic matter, such as manure or sewage, in the absence of oxygen. It can be used as a *biofuel

biome a large-scale *ecosystem, e.g. tropical rainforest

biosphere the living layer of Earth between the *lithosphere and *atmosphere

biotic living part of a *biome, made up of plant (flora) and animal (fauna) life

black gold a term used for oil, as it is regarded as such a valuable commodity

bottom-up development experts work with communities to identify their needs, offer assistance and let people have more control over their lives, often run by *non-governmental organisations

brownfield sites former industrial areas that have been developed before

C

carbon dating uses radioactive testing to find the age of rocks which contained living material

carbon footprint a calculation of the total *greenhouse gas emissions caused by a person, a country, an organisation, event or product

carbon sequestration removing carbon dioxide from the atmosphere and locking it up in biotic material

carbon sinks natural stores for carbon-containing chemical compounds, like carbon dioxide or methane

Central Business District (CBD) the heart of an urban area, often containing a high percentage of shops and offices

channel refers to the bed and banks of the river

climatologist a scientist who is an expert in climate and climate change

collision zone where two *tectonic plates collide – forming mountains like the Himalayas

communism is a system of government, based on the theories of Karl Marx, which believes in sharing wealth between all people

concordant coasts follow the ridges and valleys of the land, so the rock *strata is parallel to the coastline

connectivity how easy it is to travel or connect with other places

conservation means protecting threatened *biomes, e.g. setting up national parks or banning trade in endangered species

conservative boundary where two *tectonic plates slide past each other

constructive waves build beaches by pushing sand and pebbles further up the beach

continental crust the part of the Earth's crust that makes up land, on average 30-50 km thick

conurbation a continuous urban or built-up area, formed by merging towns or cities

convection currents transfer heat from one part of a liquid or gas to another. In the Earth's *mantle, the currents which rise from the Earth's core are strong enough to move the *tectonic plates on the Earth's surface

convergence there are two meanings: a) the coming together of *tectonic plates and b) when air streams flow to meet each other

Coriolis force a strong force created by the Earth's rotation. It can cause storms, including hurricanes

cost-benefit analysis looking at all the costs of a project, social and environmental as well as financial, and deciding whether it is worth going ahead

counter-urbanisation when people leave towns and cities to live in the countryside

D

death rate number of deaths per 1000 people per year

decentralisation shift of shopping activity and employment away from the *Central Business District (CBD)

deforestation the deliberate cutting down of forests to exploit forest resources (timber, land or minerals)

deindustrialisation decreased activity in manufacturing and closure of industries, leading to unemployment

delta a low-lying area at the mouth of a river where a river deposits so much *sediment it extends beyond the coastline

dependency ratio proportion of people below (aged 0-14) and above (over 65) normal working age. The lower the number, the greater the number of people who work and are less dependent

depopulation decline of total population of an area

deprivation lack of wealth and services. It usually means low standards of living caused by low income, poor health, and low educational qualifications

dip slope a gentle slope following the angle of rock *strata, found behind *escarpments

discharge the volume of water flowing in a river, measured in cubic metres per second

discordant coast alternates between bands of hard rocks and soft rocks, so the rock *strata is at right angles to the coast

dissipate means to reduce wave energy, which is absorbed as waves pass through, or over, sea defences

divergent plate boundary where two *tectonic plates are moving away from each other

diversification when a business (e.g. a farm) decides to sell other products or services in order to survive or grow

E

ecological debt when Earth's resources are being used up faster than Earth can replace them

ecological footprint is a calculation measured in global hectares (gha). It's the amount of land and water required to produce resources and deal with waste from each country

173

Glossary

economic liberalisation when a country's economy is given the freedom of a 'market economy', consumers and companies decide what people buy based on demand

ecosystem a localized *biome made up of living things and their non-living environment. For example a pond, a forest, a desert

ecosystem services a collective term for all of the ways humans benefit from ecosystems

emerging economies countries that have recently industrialised and are progressing towards an increased role in the world economy

energy diversification getting energy from a variety of different sources to increase *energy security

energy security having access to reliable and affordable sources of energy

enquiry the process of investigation to find an answer to a question

epicentre the point on the ground directly above the focus (centre) of an earthquake

erosion means wearing away the landscape

escarpment a continuous line of steep slopes above a gentle *dip slope, caused by the erosion of alternate *strata

evacuate when people move from a place of danger to a safer place

evaporation the changing of a liquid into vapour or gas. Some rainfall is evaporated into water vapour by the heat of the sun

F

fault large cracks caused by past tectonic movements

fertility rate average number of births per woman

fetch the length of water over which the wind has blown, affecting the size and strength of waves

fieldwork means work carried out in the outdoors

flood plain flat land around a river that gets flooded when the river overflows

focus the point of origin of an earthquake

food miles the distance food travels from the producer to the consumer. The greater the distance, the more carbon dioxide is produced by the journey

food web a complex network of overlapping food chains that connect plants and animals in *biomes

formal economy means one which is official, meets legal standards for accounts, taxes, and workers' pay and conditions

fossil fuels a natural fuel found underground, buried within sedimentary rock in the form of coal, oil or natural gas

free trade the free flow of *goods and *services, without the restriction of tariffs

friction the force which resists the movement of one surface over another

G

gentrification high-income earners move into run-down areas to be closer to their workplace, often resulting in the rehabilitation and *regeneration of the area to conform with middle class lifestyles

geographical conflict means disagreement and differences of opinion linked to the use of places and resources

geographical information systems (GIS) a form of electronic mapping that builds up maps layer by layer

geothermal heat from inside the Earth

glacial a cold period of time during which the Earth's glaciers expanded widely

global circulation model a theory that explains how the *atmosphere operates in a series of three cells each side of the Equator

global shift change in location of where manufactured goods are made, often from developed to developing countries

globalisation increased connections between countries

goods physical materials or products that are of value to us

green belt undeveloped areas of land around the edge of cities with strict planning controls

greenhouse effect the way that gases in the atmosphere trap heat from the sun. Like the glass in a greenhouse – they let heat in, but prevent most of it from escaping

greenhouse gases gases like carbon dioxide and methane that trap heat around the Earth, leading to global warming

gross domestic product (GDP) the total value of *goods and *services produced by a country in one year

groundwater flow movement of water through rocks in the ground

H

hard engineering building physical structures to deal with natural hazards, such as sea walls to stop waves

helicoidal flow a continuous corkscrew motion of water as it flows along a river channel

holistic management takes into account all social, economic and environmental costs and benefits. In coastal management this means looking at the coastline as a whole instead of an individual bay or beach

hot spot columns of heat in Earth's *mantle found in the middle of a tectonic plate

Human Development Index (HDI) a standard means of measuring human development

hydraulic action the force of water along the coast, or within a stream or river

hydrological cycle the movement of water between its different forms; gas (water vapour), liquid and solid (ice) forms. It is also known as the water cycle

hyper-urbanisation rapid growth of urban areas

I

Index of Multiple Deprivation (IMD) means of showing how deprived some areas are

indigenous peoples are the original people of a region. Some indigenous groups still lead traditional lifestyles, e.g. a tribal system, hunting for food

industrialisation where a mainly agricultural society changes and begins to depend on manufacturing industries instead

infant mortality number of children per 1000 live births who die before their first birthday

infiltration the soaking of rainwater into the ground

informal economy means an unofficial economy, where no records are kept. People in the informal economy have no contracts or employment rights

infrastructure the basic services needed for an industrial country to operate e.g. roads, railways, power and water supplies, waste disposal, schools, hospitals, telephones and communication services

interception zone the capture of rainwater by leaves and branches. Some *evaporates again and the rest drips from the leaves to the soil

interglacial a long period of warmer conditions between *glacials

interlocking spurs hills that stick out on alternate sides of a V-shaped valley, like the teeth of a zip

intermediate technology uses low-tech solutions using local materials, labour and expertise to solve problems

Inter-Tropical Convergence Zone (ITCZ) a narrow zone of low pressure near the Equator where northern and southern air masses converge

invasive species (or alien species) is a plant, animal or disease introduced from one area to another which causes ecosystem damage

irrigation is the artificial watering of land that allows farming to take place

J

jet streams high level winds at around 6-10km that blow across the Atlantic towards the UK

joints small and usually vertical cracks found in many rocks

Glossary

L

lagoon a bay totally or partially enclosed by a *spit, *bar or reef running across its entrance

landslide a rapid *mass movement of rock fragments and soil under the influence of gravity

latitude how far north or south a location on the Earth's surface is from the Equator, measured in degrees

lava melted rock that erupts from a volcano

lava flows *lava flows at different speeds, depending on what it is made of. Lava flows are normally very slow and not hazardous but, when mixed with water, can flow very fast and be dangerous

level of development means a country's wealth (measured by its GDP), and its social and political progress (e.g. its education, health care or democratic process in which everyone can vote freely)

life expectancy average number of years a person can expect to live

lithosphere the uppermost layer of the Earth. It is cool and brittle. It includes the very top of the *mantle and, above this, the crust

M

magma melted rock below the Earth's surface. When it reaches the surface it is called *lava

magnitude of an earthquake (how much the ground shakes), an expression of the total energy released

mantle the middle layer of the Earth. It lies between the crust and the core and is about 2900 km thick. Its outer layer is the *asthenosphere. Below the asthenosphere it consists mainly of solid rock

mass movement the movement of material downslope, such as rock falls, *landslides or cliff collapse

maternal mortality number of mothers per 100 000 who die in childbirth

megacity a many centered, multi-city urban area of more than 10 million people. A megacity is sometimes formed from several cities merging together

middle course the journey of a river from its source in hills or mountains to mouth is sometimes called the course of the river. The course of a river can be divided into three main sections a) upper course b) middle course and c) lower course

migration movement of people from one place to another

Milankovitch cycles the three long-term cycles in the Earth's orbit around the sun. Milankovitch's theory is that *glacials happen when the three cycles match up in a certain way

mudflats flat coastal areas formed when mud is deposited by rivers and coasts

multicultural a variety of different cultures or ethnic groups within a society

multiplier effect when people or businesses move to an area and invest money on housing and services, which in turn creates more jobs and attracts more people

N

natural increase the birth rate minus the death rate for a place. It is normally given as a % of the total population

natural resources are materials found in the environment that are used by humans, including land, water, fossil fuels, rocks and minerals and biological resources like timber and fish

net primary productivity (NPP) a measure of how much new plant and animal growth is added to a biome each year

non-governmental organisation (NGO) NGOs work to make life better, especially for the poor. Oxfam, the Red Cross and Greenpeace are all NGOs

non-renewable energy sources that are finite and will eventually run out, such as oil and gas

northern powerhouse a major core region of cities (with a similar population to London) that has the potential to drive the economy of northern England

nutrient cycle nutrients move between the biomass, litter and soil as part of a continuous cycle which keeps both plants and soil healthy

O

ocean currents permanent or semi-permanent large-scale horizontal movements of the ocean waters

oceanic crust the part of the Earth's crust which is under the oceans, usually 6-8 km thick

Organisation of Petroleum Exporting Countries (OPEC) established to regulate the global oil market, stabilize prices and ensure a fair return for its 12 member states who supply 45% of the world's oil

outsourcing using people in other countries to provide services if they can do so more cheaply e.g. call centres

ox-bow lake a lake formed when a loop in a river is cut off by floods

P

Pangea a supercontinent consisting of the whole land area of the globe before it was split up by continental drift

peak oil the theoretical point at which half of the known reserves of oil in the world have been used

plate boundaries where *tectonic plates meet. There are three kinds of boundary a) *divergent – when two plates move apart b) *convergent – when two plates collide c) conservative – when two plates slide past one another

plumes upwelling of molten rock through the *asthenosphere to the *lithosphere

plunging waves typically tall and close together, created by strong winds

population density the average number of people in a given area, expressed as people per km²

population structure the number of each sex in each age group (e.g. 10-14), usually displayed in a population pyramid diagram

poverty line the minimum level of income required to meet a person's basic needs (US$1.25)

predict saying that something will happen in the future. A scientific prediction is based on statistical evidence

prevailing winds the most frequent direction the wind blows in a certain area

primary effects the direct impacts of an event, usually occurring instantly

primary products raw materials

Purchasing Power Parity (PPP) shows what you can buy in each country, now used to measure *GDP

pyroclasts fragments of volcanic material that is thrown out during explosive eruptions

Q

quality of life a measure of how 'wealthy' people are, but measured using criteria such as housing, employment and environmental factors, rather than income

Quaternary the last 2.6 million years, during which there have been many *glacials

R

radioactive decay atoms of unstable elements release particles from their nuclei and give off heat

rebranded a change of image

recurved hooked

regeneration means re-developing former industrial areas or housing to improve them

renewable a resource that does not run out and can be restored, such as wind or solar

re-urbanisation when people who used to live in the city and then moved out to the country or to a suburb, move back to live in the city

Richter scale a scale for measuring the magnitude of earthquakes

rock outcrop a large mass of rock that stands above the surface of the ground

rockfalls a form of *mass movement where fragments of rock fall freely from a cliff face

Glossary

rural-urban fringe the area where a town or city meets the countryside

rural-urban migration the movement of people from the countryside to the cities, normally to escape from poverty and to search for work

S

Saffir-Simpson Hurricane Scale a scale that classifies hurricanes into five different categories according to their wind strength

salt marsh salt-tolerant vegetation growing on mud flats in bays or estuaries. These plants trap sediments which gradually raise the height of the marsh. Eventually it becomes part of the coast land

saltation the bouncing of material from and along a river bed or a land surface

sand dune onshore winds blow sand inland, forming a hill or ridge of sand parallel to the shoreline

saturated soil is saturated when the water table has come to the surface. The water then flows overland

scree angular rock pieces created by freeze-thaw weathering

secondary effects the indirect impacts of an event, usually occurring in the hours, weeks, months or years after the event

secondary products manufactured goods

sediment material such as sand or clay that is transported by rivers

seismometer a machine for recording and measuring an earthquake using the *Richter scale

services functions that satisfy our needs

Shoreline Management Plan (SMP) this is an approach which builds on knowledge of the coastal environment and takes account of the wide range of public interest to avoid piecemeal attempts to protect one area at the expense of another

slope processes cause *mass movement or *soil creep

socio-economics how economic activity affects, and is shaped by, social processes

soft engineering involves adapting to natural hazards and working with nature to limit damage

soil creep the slow gradual movement downslope of soil, *scree or glacier ice

solution chemicals dissolved in water, invisible to the eye

spatial means 'relating to space' e.g. the spatial growth of a city means how much extra space it takes up as it grows

spit a ridge of sand running away from the coast, usually with a curved seaward end

stakeholders a person with an interest or concern in something, such as those who are likely to be affected by natural hazards

storm hydrograph a graph which shows the change in both rainfall and discharge from a river following a storm

storm surge a rapid rise in the level of the sea caused by low pressure and strong winds

strata distinctive layers of rock

stratosphere the layer of air 10-50km above the Earth's surface. It is above the cloudy layer we live in, the troposphere

strip mining (or open-pit, opencast or surface mining) involves digging large holes in the ground to extract ores and minerals that are close to the surface

studentification communities benefit from local universities which provide employment opportunities and a large student population which can regenerate pubs, shops and buy-to-let properties

sub-aerial processes occurring on land, at the Earth's surface, as opposed to underwater or underground

subduction describes *oceanic crust sinking into the *mantle at a convergent *plate boundary. As the crust subducts, it melts back into the mantle

subsistence farming where farmers grow food to feed their families, rather than to sell

suburbanisation the movement of people from the inner suburbs to the outer suburbs

surface run-off rainwater that runs across the surface of the ground and drains into the river

suspension tiny particles of *sediment dispersed in water

Sustainable Development defined by the Brundtland Commission as that which 'meets the needs of the present without compromising the ability of future generations to meet their own needs'

sustainable management meeting the needs of people now and in the future, and limiting harm to the environment

T

tar sands sediment that is mixed with oil, can be mined to extract oil to be used as fuel

tectonic hazards natural events caused by movement of the Earth's plates that affect people and property

tectonic plate the Earth's surface is broken into large pieces, like a cracked eggshell. The pieces are called tectonic plates, or just plates

terms of trade means the value of a country's exports relative to that of its imports

thalweg the line of the fastest flow along the course of a river

thermal expansion as a result of heating, expansion occurs. When sea water warms up, it expands

throughflow the flow of rainwater sideways through the soil, towards the river

till sediment deposited by melting glaciers or ice sheets

top-down development when decision-making about the development of a place is done by governments or large companies

topography the study of the shape and features of the surface of the Earth. Can refer to the shapes and features themselves, or a depiction on a map

traction force that rolls or drags large stones along a river bed

transnational companies (TNCs) those which operate across more than one country

transpire when plants lose water vapour, mainly through pores in their leaves

tsunami earthquakes beneath the sea bed generate huge waves that travel up to 900km/h

U

urbanisation means a rise in the percentage of people living in urban areas, compared to rural areas

V

velocity the speed of a river, measured in metres per second

Volcanic Explosivity Index (VEI) measures the explosiveness of volcanic eruptions on a scale of 1 to 8

W

water table the upper limit of saturated rock below the ground

wave-cut platform the flat rocky area left behind when waves erode a cliff away

weathering the physical, chemical or biological breakdown of solid rock by the action of weather (e.g. frost, rain) or plants

wildfire uncontrolled burning though forest, grassland or scrub. Such fires can 'jump' roads and rivers and travel at high speed

world cities trade and invest globally e.g. London and New York